Polymer Composite Materials – Interface Phenomena & Processes

SOLID MECHANICS AND ITS APPLICATIONS
Volume 90

Aims and Scope of the Series

The fundamental questions arising in mechanics are: *Why?*, *How?*, and *How much?* The aim of this series is to provide lucid accounts written by authoritative researchers giving vision and insight in answering these questions on the subject of mechanics as it relates to solids.

The scope of the series covers the entire spectrum of solid mechanics. Thus it includes the foundation of mechanics; variational formulations; computational mechanics; statics, kinematics and dynamics of rigid and elastic bodies: vibrations of solids and structures; dynamical systems and chaos; the theories of elasticity, plasticity and viscoelasticity; composite materials; rods, beams, shells and membranes; structural control and stability; soils, rocks and geomechanics; fracture; tribology; experimental mechanics; biomechanics and machine design.

The median level of presentation is the first year graduate student. Some texts are monographs defining the current state of the field; others are accessible to final year undergraduates; but essentially the emphasis is on readability and clarity.

Polymer Composite Materials – Interface Phenomena & Processes

by

YATCHKO IVANOV

VALERII CHESHKOV

and

MARGARITA NATOVA

Bulgarian Academy of Sciences,
Central Laboratory of Physico-Chemical Mechanics,
Sofia, Bulgaria

KLUWER ACADEMIC PUBLISHERS
DORDRECHT / BOSTON / LONDON

A C.I.P. Catalogue record for this book is available from the Library of Congress.

ISBN-13: 978-1-4020-0386-8 e-ISBN-13: 978-94-010-9664-5
DOI: 10.1007/978-94-010-9664-5

Published by Kluwer Academic Publishers,
P.O. Box 17, 3300 AA Dordrecht, The Netherlands.

Sold and distributed in North, Central and South America
by Kluwer Academic Publishers,
101 Philip Drive, Norwell, MA 02061, U.S.A.

In all other countries, sold and distributed
by Kluwer Academic Publishers,
P.O. Box 322, 3300 AH Dordrecht, The Netherlands.

Printed on acid-free paper

Dedicated to our children:
Galina
Sergey, Cvetan
Tsvetomira, Michaela, Martina

Contents

Foreword

1. OBJECTIVES AND SCOPE

Polymer composite materials are widely used in practice and are an object of numerous, either scientific or scientific-applied, studies. Note that the prospects of study of these materials are so promising that it is difficult to name even a single branch of industry where composites have not been applied. Regarding structure and properties, polymer composites comprise such materials as glass and carbon plastics, polymer concretes and concrete polymers, varnish-dye materials, wood-polymer materials, reinforced elastomers, etc. Their significance is due to a number of useful properties that pure, unfilled polymers do not possess. The purposeful introduction of fillers into polymers provides for the change of material properties, thus gaining a definite profit. Moreover, it is worth noting that polymer filling provides new possibilities for significant decrease in material cost.

The practically infinite set of fillers enables one to vary in quite wide ranges the polymer properties and those of polymer-based composite materials. The excellent physico-mechanical properties of polymer composites, for an arbitrary low material density, seem promising for their highly effective use in industry and instrument design, in car industry, civil engineering, light industry, agriculture and in other areas. For each specific case, it is possible to choose or develop polymer composite that meets the necessary requirements, i.e. that have properties in accordance with its eventual application.

It is important to note that various industrial wastes, whose secondary application has not yet been technically or economically assessed, can be used as polymer fillers, resulting in a definitely positive ecology effect. These are, most of all, such plentiful wastes as phosphogypsum, various slags, ash from thermal power stations, etc. that considerably pollute the environment.

When tackling the prospects of polymer composite development, we should note that a number of technical-economical and technological difficulties arise in the process of material filling. Furthermore, there are some problems linked with the nature of polymer composite material. These problems are due to the fact that the fillers introduce interfaces where various complex physical, chemical and physico-chemical processes take place. They start at the moment of mixing the composite components and proceed during the formation of the material structure. Besides, some of the processes do not cease even after final material structure formation and could significantly affect further polymer destruction.

Regarding the numerous reference data, we may conclude that the interface phenomena that occur in polymer composite materials significantly affect the formation of material structure and properties. Hence, the study of the development of interface processes and possibilities for their control is most significant. Such a profound study is necessary to choose the appropriate filler for polymers and to take into consideration both the properties of the separate composition components and the desired properties of the polymer composite. Furthermore, the analysis yields the necessity of deriving the laws of polymer composite formation, considering a wide range of methods and their complex use in one and the same polymer system.

Such a set-up faces a number of difficulties that accompany the use of reference data. This is due to the fact that often data for one and the same composite material, but obtained by different authors, can not be compared and generalized. There are various reasons for such a mismatch, for example insufficient characterization of the initial composite components, use of substances with different degrees of purity, maintenance of different conditions when combining and processing separate composite components, different processing conditions, etc. These factors can affect either the development of surface interface processes or the formation of structure and properties of the polymer composites. The analysis and generalization of data are significantly hampered, since a number of authors reduce their studies to one or another aspect of the design of effective polymer composites. For instance, most researchers study separately either adsorption phenomena or rheology of composites, composite processing, composite properties, etc.

We consider that a complete idea on the composite material structure formation can be obtained only by using a complex versatile approach, consisting of several successive stages. The first stage should cover the choice and characterization of the initial components of the polymer composites. The next stage should consist in the study of the processes that take place during component combination. Then, processes of the material structure formation and properties of the polymer composites should be analyzed. The study should be completed by a characterization of the composite and by establishing an interrelation between the results of the each stage. Such an approach is the basis of the present monograph.

A further basic idea developed in our study is that, for the most part, interface phenomena occurring in the systems "filler/monomer (oligomer or polymer)", should be considered as surface acid-base interactions. Such a necessity is verified by the fact that practically all fillers are either solid acids or solid bases, while the most part of monomers, oligomers and polymers can display basic or acidic behaviour. Specific examples illustrate the prospects and abilities that such a complex set-up provides for the control of structure formation, as well as for the design of new and improved composite materials.

The principal objective of the book is to close the gap between current knowledge and the available publications by emphasizing composite material formation, interface phenomena and processes. Consistently employing the ideas of material science, the monograph combines a thorough analysis of recent results on acid-base adsorption interactions in filled polymers with classical studies of filled composites. This creates a coherent framework that unifies the subject and makes the available literature more accessible, especially for those wishing to extend their knowledge on particular composite material systems.

An additional objective is to analyze the adsorption interactions between polymers and the solid phase. We also provide an overview of the latest advances in fields, such as structural physico-chemical mechanics, composite rheologycal behaviour and property formation in the presence of solid acids or bases. Our language is intended to be as clear and understandable as possible and we pay attention to both chemistry and physics needed to understand various phenomena that occur in composite materials. However, we aim at reaching a balance between exact presentation, intuitive grasp of new ideas and creative application of concepts throughout the text.

Since the text incorporates essential physical, chemical and physico-chemical fundamentals, it is largely self contained and appropriate for readers with interests in a wide range of disciplines of material science. It is both a monograph for academic and industrial researchers and a textbook for teaching advanced courses in subjects such as civil and mechanical engineering, physics and applied mechanics. The coverage of the book is rather broad and some of the chapters have a survey character, while others set forth details. This enables the researcher or the instructor to select those sections that suit their goals. In fact, the book is based on authors'results that have been obtained after a long collaboration in the study of organic composite materials.

2. ORGANIZATION OF THE BOOK

The book is divided into five chapters.

Chapter 1 is itself divided into 4 parts and sets forth the basis of the ***acid-base*** approach for tackling the process of polymer filling. The discussion of the acid-base properties of all components of the polymer composite materials – both fillers and polymers, is deep and well-founded. Special attention is paid to the analysis of the nature of acidic and basic sites of the inorganic substances that can serve as polymer fillers. A survey is made of the methods used for the determination of the nature, strength and concentration of surface acidic and basic sites of the fillers. A wide class of disperse materials is discussed, too. It is shown that most of materials used as fillers of polymer composite materials can be classified as solid acids. The change of the acid-base properties of the filler surface, occurring during filler modification via thermal, chemical or radiation treatment, is thoroughly considered. Polymers are classified into four groups – acidic, basic, bifunctional and neutral ones.

Chapter 2 is composed of 4 parts and presents the theoretical aspects of ***adsorption processes*** that develop during polymer filling from an acid-base point of view. The formation of adsorption species between fillers (solid acids or bases), on the one hand, and between basic or acidic in character polymers, on the other, are described in detail. Attention is also paid to the methods used for the determination of acid-base interaction between phases, while examples of typical acid-base interactions are given.

Chapter 3 tackles the principal ***rheological*** properties of a polymer composite and changes in its rheological behaviour. These changes are due to a number of factors – interaction between phases, presence of fillers that contain acidic or basic sites, filler amount and type. The results of the study of the rheology behaviour and properties of polymer composites, containing acidic or basic fillers, are presented in detail. The effect of the interaction between phases on viscosity, viscoelastic characteristics and thixotropy of polymer compositions is outlined. Special attention is paid to the variation of these properties by performing a surface modification of disperse fillers. Possible ways are proposed for regulating and controlling the changes of the rheological properties as well as of the technological parameters during processing of filled polymer composites.

Chapter 4 covers the ***basic parameters*** and peculiarities of polymer composite formation from solutions and melts of monomers, oligomers and polymers, and in the presence of fillers (solid acids or bases). Special attention is paid to the ***catalytic*** effect of acid and base filler surface sites on the process of polymerization and polycondensation. A correlation between heterogeneous acid-base catalysis and homogeneous catalysis in polymer formation is scrutinized. Possibilities of controlling the processes during the time of preparation of filled polymer composites by including acidic or basic in character fillers are thoroughly discussed.

The last part of the book, Chapter 5, treats ***structures*** and ***properties*** of polymer composite materials that contain fillers – solid acids or bases. Advantages and prospects for further development and improvement of polymer composite materials with active fillers – solid acids or bases, are underlined, comparing their properties with the physico-mechanical and operating characteristics of polymers, filled with inactive fillers. Possibilities of preparing composite materials of high radiation protection are specially outlined. A separate section deals with the ways of a filler acid-base modification, aiming at the improvement of polymer composites properties. The concluding section discusses pending problems and future prospects for employing the acid-base approach to the design and analysis of composite materials and for modeling and predicting their properties. Prospects for further studies and possible application of filled polymers are outlined, too. The monograph is based on the concept that the acid–base interactions between phases are crucial for polymer composite formation.

Acknowledgements

We would like to express our sincere appreciation to the Series Editor, Professor Graham Gladwell, for his invaluable help in editing the wording and contents of the text. We wish to express our cordial thanks to Associate Professor Robert Kazandjiev, for his creative participation in the preparation of the book for publication in English, and to Mrs. V. Ivanova, Mrs. D. Simenska, Mr. D. Tomov, and other members of the staff of the Central Laboratory of Physico-Chemical Mechanics, for their technical assistance in preparing the manuscript and figures in camera-ready form.

Finally we thank our families for their patience and support during the long preparation of the book.

Chapter 1

ACID-BASE PROPERTIES OF THE COMPONENTS OF POLYMER COMPOSITE MATERIALS

1. GENERAL PRINCIPLES

The set of substances that are used or may be used as polymer fillers is quite large. It comprises synthetic or natural inorganic and organic compounds, such as powdered, fibrillar or fabric ones. First of all, powdered fillers, being finely or coarsely dispersed, may be divided into two groups: ***inorganic*** compounds, such as oxides, metals, carbonates, phosphates, sulphates, talc, clays, mica, blacks, graphite, perlite, etc., and ***organic*** compounds - polystyrene, rubber, wood particles, etc. Secondly, a rich variety of fibrillar fillers are employed- ***inorganic*** fibers, such as glass, carbon and basalt fibers, and ***organic*** fibers - polyamide, aramid and viscose fibers. Finally, glass and textile fabrics are used as fillers.

We shall consider in our further study, most of all, disperse inorganic fillers, since they are widely used for the preparation of filled composite materials. Moreover, they can be taken as an initial basis for a comparison if the polymer composite system is planned to be further complicated.

As noted, filler introduction not only yields a decrease in the polymer material *price*, but also decreases material *contraction* and the tendency to *destruction* and *ageing*; it increases *heat-*, *water-*, *fire* and *chemical resistance* of the material and improves material *thermal stability*, *electro-* and *thermal insulating* properties; it governs *sorption* development, changes polymer gas- and vapour *permeability*, improves material *processability*, etc.; finally, the fillers may affect the *kinetics* and *mechanisms* of the formation of a definite class of polymers. We will discuss all these effects of the fillers on the polymer composite.

The set of fillers for polymers significantly widens when the filler surface is modified by some treatment - chemical, thermal and mechanochemical, plasma treatment or exposure to irradiation. Such a modification most often yields activation of the filler surface, improvement of the compatibility between the filler and the monomer, oligomer or polymer, strengthening of the filler-polymer adhesion, etc. Note that there is a large number of data on the combination of the fillers and polymers, but filling theory often falls behind filling practice.

Even though the introduction of fillers generally improves the polymer characteristics, an improvement in one property might lead to the worsening of another. One must compromise to obtain reasonable overall characteristics of the final polymer composite material.

The characteristics of polymer composite materials depend not only on the nature and properties of their components, but also on the methods of material *processing*. Various methods are employed to join the filler and the polymer: filler mixing with a polymer melt, filler introduction into a polymer solution, filler mixing with powdered polymer, filler introduction into a monomer or oligomer, filler formation within a polymer medium by using a thermal decomposition of some compounds, etc. Even though the technology of mixing considerably affects polymer composite properties, there is a lack of comparative data on these effects in practice. To illustrate the significance of the correct choice of a method of filler introduction we present a comparison of some known data on the development of interphase processes which depend on the conditions of polymer composite formation.

Various processes may take place when a certain filler is introduced into a polymer. They may have a significant influence on the properties of the final material. Note that a change of the proportion between the high- and the low molecular fractions within the polymer volume may occur as a result of adsorption. Moreover, there may be significant adsorption of one of the components of the multicomponent system, and the filler surface may become physically and chemically bonded to the solvent and to the monomer, oligomer or polymer. Next, specific arrangements and orientations of the molecules of the monomer, oligomer or polymer may occur near the filler surface. Finally, filler particles may either serve as steric obstacles to crack propagation within the polymer composites, or be stress concentrators and potential crack initiators. All these are specific and illustrative examples of the phenomena that may occur in filled polymer composites.

Filler effects on a polymer can be classified as structural, kinetic and thermodynamic. ***Structural*** effects are displayed by the change of the parameters of the supermolecular formations, by the macromolecular crosslinking density within the spatial lattice, by a change of the proportion between the ordered and amorphous polymer areas, and by change of the degree of polymer crystallinity. ***Kinetic*** effects are usually linked with the filler capability to influence either the mobility of different elements of the macromolecular chain, or the mobility of the whole chain. Hence, the filler is characterised with a capacity to control the spectrum of polymer relaxation times. In addition, filler ability to affect the rate of adsorption and desorption of molecules of monomer, oligomer and polymer on and from the filler surface should also be considered as a kinetic effect. Such a rate factor determines the duration of the contact between the adsorbing molecules and filler surface, i.e. the time of adsorption interaction. Fillers may also influence polymer ***thermodynamic*** characteristics, significantly affecting the properties of the polymer layer that is closest to the filler surface.

In a detailed analysis of the known data, it seems useful to outline one more filler effect on polymer composite systems, i.e. a ***chemical*** one. The latter may affect the reaction ability of oligomers and monomers, or somehow catalyze polymerization and polycondensation processes. Note that as compared to a unfilled polymer, the filler may selectively accelerate one reaction relative to other possible ways of polymerization and polycondensation, and also affect the chemical constitution of a filled polymer.

Different filling effects may occur in different polymer fabrications; they depend on the specific polymer composition, on the conditions of its preparation and on further treatment and processing.

Filler activity in polymer composites depends on a number of factors: filler particle shape and size, particle size distribution, specific surface and porosity, packing structure of the filler into the polymer, and its physico-mechanical properties. While not being decisive for a filler activity and its effects on the polymer, these factors play a major role in the formation of polymer composite materials. Hence, their importance is crucial in determing the state of the filler surface, i.e. filler chemistry and physico-chemistry. Filler surface chemical and physico-chemical properties are characterized by sites or groups, existing on the filler surface, and being able of interacting with one or another monomer, oligomer or polymer groups. Active surface sites are imperfections of the filler crystalline structure, admixture atoms and molecules, co-ordination unsaturated ions, hydroxyl groups, coordination-bonded water molecules, etc. Their activity depends on the properties of the specific adsorbate - monomer, oligomer or polymer. The interactions between them may vary from a very weak one (of a van der Waals type) to a very strong one (formation of chemical links). Some aspects of the various interphase processes that take place in filler/monomer (oligomer, polymer) systems are discussed in a number of papers [1-9].

The variety of active sites that may exist on the filler surfaces necessitates site analysis, and this seems to be a very important study. Filler surface properties have still not been entirely characterized, and no attention has been paid to their sensitivity to various effects. For instance, even an usual impurity of the filler surface owing to polluted environment (adsorption of moisture, carbon dioxide, etc.) may yield a sharp change of surface properties and, hence, may modify a filler influence on the polymer. Yet, the effect of filler surface chemistry and physico-chemistry is not entirely known even in those cases where one and the same polymer composition is considered.

In most cases, interphase processes that take place during polymer filling can be tackled from an acid-base point of view. This is so since all compounds used as fillers for polymers may be treated as solid acids or bases. Furthermore, monomers, oligomers and polymers may be considered as compounds that are able of displaying acidic or basic properties. Thus, it proves useful to study the nature of a filler acidity (or basicity), a surface site strength and concentration, as well as the corresponding monomer, oligomer and polymer properties, joining them to the appropriate fillers.

2. NATURE, STRENGTH AND CONCENTRATION OF ACIDIC AND BASIC FILLER SITES

Most of the studies that deal with polymer filling establish that the filler is either acid or base, but they do not submit data on a filler acidity or basicity. This is typical even for researches that discuss the acid-base interactions in filled polymers and how these interactions affect the properties of the composite materials [10, 11].

For now, most detailed and exact description of the acid-base properties are given only for such inorganic compounds that are widely used as catalysts in petrochemistry and in the organic synthesis [12]. Catalysis on solid acids and to a certain extent on solid bases is applied in a number of reactions, such as cracking, isomerization, alkylation, dehydration, etc. Despite the specificity of these catalytic reactions, such trends of science and practice may provide data and methods that would enable one to characterize the fillers for polymers from an acid-base point of view. Moreover, a number of catalysts - solid acids or bases, may be used as polymer fillers.

Brønsted and Lewis define a solid acid as a compound that is able of donating protons or accepting an electron pair, while a solid base is a compound that is a proton acceptor or an electron pair supplier [12].

In the donor-acceptor theory of Brønsted and Lewis [12-14], the following compounds may be considered as solid acids - silicon, chromium, ferric and other oxides, sulphates, silicates, phosphates and chlorides. Solid bases include calcium, magnesium, zinc and other oxides, carbonates, as well as certain complex oxides. Aluminium and titanium oxides and aluminosilicates can have a bifunctional character, i.e. they may display either acidic and basic properties.

Surface cations that are coordination-unsaturated, can play the role of acidic sites of a Lewis type. Owing to incomplete electron orbitals, they may display electron accepting character. Brønsted acidity, i.e. a proton donating acidity, is determined by various types of hydroxyl groups that exist on the solid surface. These, too, may display basic properties under specific conditions, but the surface basicity is generally related to surface anions that have unpaired electron pairs.

As for an acidity or basicity of solids, we consider not only the nature of their acidic or basic sites located on the surface, but also a site strength and concentration. The strength of the acidic sites is specified by the ability of the solid surface to transform the adsorbed basic molecules into their conjugated acidic forms. Conversely, the strength of the basic sites can be specified by the ability of the solid surface to transform the adsorbed acidic molecules into their conjugated basic forms. Depending on the specific chemical reaction and employing solid acids or bases as catalysts, one of these three parameters - site nature, strength or concentration, is the factor that prevails over the chemical process [12]. Site strength and concentration that characterize acidity or basicity vary in very wide ranges, even for one and the same solid compound. Generally, the surface acidity or basicity is not a constant quantity for a given inorganic surface. Both acidity and basicity depend on various factors - methods and conditions of the solid compound fabrication, methods of additional treatment - thermal, chemical, radiation one, etc. This explains to a large extent the different, sometimes contradictory, data on the acid-base properties of one and the same solid compound or of one and the same composite material. What is said holds true for those cases where solid acids or bases are used as fillers for polymer materials. Hence, it is not only useful, but also necessary to characterize fillers from an acid-base point of view, before using them in each specific case.

We should note that surface properties of solid acids are studied more profoundly than those of solid bases; solid acids are more often used as effective catalysts in various organic reactions. The same is the case with polymer filling, too, since compounds that display acidic surface properties are most often used as fillers. This, however, can be easily explained: for the most part, monomers, oligomers and polymers may be treated as compounds that display basic properties. Regarding the properties of polymer composite materials, it is simultaneously established that the best results are obtained when acidic fillers are used in basic monomers, oligomers and polymers and *vice versa* [10]. Hence, the idea that the acid-base interphase interactions play a decisive role during the formation of filled polymers becomes a basis of further considerations. This also explains the tendency to combine components with opposite acid-base nature during polymer filling, and why solid acids are most often used as fillers.

The set of methods that are used to characterize the acidic and basic properties of inorganic surfaces is large: infrared and ultraviolet spectroscopy, nuclear magnetic and electronic paramagnetic resonance, titration, calorimetry, etc. [15-21]. Recently, the method of inverse gas chromatography [22] has been widely used for a determination of the potential of donor-acceptor interactions characteristic of polymer system components.

All these methods employ adsorption of special, low molecular acidic or basic compounds termed ***sondes***, since surface sites display their character only during the interactions and the direct study of whether an isolated site has an acidic or basic character has no sense. As mentioned above, three parameters are used for the characterization of the acidity or basicity of solid compounds: site nature (acidic or basic), strength and concentration.

The type of the surface sites is most often determined by using infrared spectroscopy; it produces irrefutable results. Thus, changes of the spectral characteristics of either the adsorbed molecules (sondes) [23-25] or of the solid body itself [26, 27] can be established and examined. Note that the former method, i. e. study of sonde spectral characteristics, is often used. Due to the large number of adsorbates - sondes, either acidic (Brønsted and/or Lewis) or basic surface sites of different strength can be identified.

Various compounds, sondes, are used to identify the acidic sites: ammonia [28, 29], pyridine [30, 31], butylamine [32, 33], acetonitrile [23, 34], benzonitrile [35], carbon monoxide [36, 37]. Note that ammonia and pyridine are very popular as site identifiers, since their infrared spectra enable one to determine not only protonic, but also aprotonic adsorbent sites. Absorption bands of the ionized forms of ammonia and pyridine are used to identify Brønsted acidic sites that protonize base molecules. Spectrum absorption bands of coordination-bonded molecules of ammonia and pyridine prove the existence of Lewis acidic sites on the surface of a solid compound. Infrared spectra of adsorbed nitriles or carbon monoxide help in the identification of Lewis acidic sites, while those of adsorbed butylamine lead to the identification of Brønsted acidic sites. Compounds such as pyrrole [38] and chloroform [39], are used as sondes for the identification of basic surface sites. The existence of basic surface sites on the adsorbent surface is proved by the shift of the valence absorption band ν_{CH} within the adsorbate spectrum. For the present, we may conclude that the problem of establishing the existence of acidic and basic sites on the solid surface has been satisfactory solved.

Significantly more complex is the problem of determining the strength of acidic and/or basic sites. Among the popular methods of site strength determination, the following techniques are worth noting: adsorption of coloured indicators from a neutral solvent; successive adsorption and desorption of gaseous acids and bases [40]; calorimetry [41]; spectroscopy [42]; gas chromatography [22], thermogravimetry and DSC [43]. Note that a specific strength scale is obtained, depending on the methods used. It is also difficult to compare results obtained by employing different methods. In addition, comparison between results grows more difficult, since the solid compounds are characterised in their most part with acidic or basic sites of different strength, i.e. the establishment of site distribution with respect to the strength is also needed.

Mostly popular here is the method of adsorption of coloured indicators (basic or acidic ones) from a neutral solvent. This method was proposed by Walling in 1950 [44]. It uses a set of indicators that have different acidic or basic strength and change their colour, depending on the strength of the adsorbent surface sites. Besides, the site strength is given as a dissociation constant pK_a or as an Hammett function H_o,, i.e. in a form that is employed for the determination of a solution acidic or basic strength. The advantages and disadvantages of the method are discussed in detail in [45]. The site strength scale is in units of pK_a or H_o and proves to be informative enough [24].

When base indicators interact with protonic sites, i.e. a proton transfer from the solid surface to the adsorbed molecules takes place, the acidic strength is expressed by means of Hammett functions:

$$H_o = pK_a + \log[B]/[BH^+] \qquad (1)$$

or

$$H_o = -\log a_{H+} \cdot f_B / f_{BH+}, \qquad (2)$$

where a_{H+} is a proton activity, f_B and f_{BH+} are the activity coefficients of the neutral base and conjugated acid, respectively, while *[B]* and *[BH⁺]* are the corresponding concentrations.

If the solid compound surface contains electron acceptors, i.e. aprotonic sites, then the acidic strength is expressed by a Hammet function H_o that has the form

$$H_o = pK_a + \log\{[B]/[AB]\}, \qquad (3)$$

or

$$H_o = -\log a_A (f_B / f_{AB}), \qquad (4)$$

where a_A is the activity of the electron pair acceptor, *[B]* and *[AB]* are the concentrations of the neutral base and of the compound obtained as a result of an addition of a base to a Lewis acidic site, or it is an electron pair acceptor *A*, while f_B and f_{AB} are the corresponding activity coefficients.

The ability of a solid compound surface to donate an electron pair to the molecule of the adsorbate (acid) characterizes the strength of the basic sites. For a reaction between an acidic indicator *AH* and a solid base $\overline{B}$

$$AH + \overline{B} \Leftrightarrow A^{-} + \overline{B}H^{+}, \tag{5}$$

the strength of the basic sites is determined as

$$H_0 = pK_a + \log [A^{-}] / [AH] \tag{6}$$

where *[AH]* is the concentration of the acidic form of the indicator (adsorbate) and $[A^{-}]$ is the concentration of their basic form.

The change of the colour of the indicator during adsorption is usually fixed visually, but for cases where the adsorbents are coloured, a register of the colour change is performed by using some physical method, e.g. ultraviolet spectroscopy [12].

The range of the indicator strength is wide. It enables one to determine not only the weakest but the strongest acidic and basic surface sites of solid fillers, i.e. from a $pK_a = +6.8$ for a neutral red (for weak acidic sites) to $pK_a = -8.2$ for an anthraquinone (for strong acidic sites) and from $pK_a = +7.2$ for a bromothymol (for weak basic sites) to $pK_a = +26.5$ for a 4-chloroaniline (for strong basic sites). Due to the non-equilibrium character of indicator adsorption, the strength of acidic and basic sites obtained through this method is not a thermodynamic quantity. Despite that, the scale for pK_a provides for a real estimation of the surface site strength and enables one to compare solid acids and bases. Note that different acidic sites could be differentiated in nature by using this method if aromatic alcohols are used as indicators, since they interact only with protonic surface sites.

Table 1 presents data obtained in our laboratory on the strength pK_a of the acidic sites of the inorganic compounds that we use in our study and in which purity is *analytical grade* [46]. The acidic site strength is determined by using the method of *n*-butylamine titration of fillers that are suspended in an anhydrous organic solvent (cyclohexane). Titration is performed in the presence of indicators (weak nitrogen containing bases): neutral red with $pK_a = +6.8$, methyl red with $pK_a = +4.8$, *p*-dimethylaminoazobenzene with $pK_a = +3.3$, dycinnamalacetone with $pK_a = -3.0$, benzophenone with $pK_a = -5.6$, etc.

Table 1. Strength of the filler acid sites

Filler	pK_a
TiO_2 (rutile)	+4.8 ÷ +6.8
SiO_2 (quartz)	+3.3
γ-Al_2O_3	Acid sites lack
$BaSO_4$	+4.8
$CaSO_4.2H_2O$	+4.8
$Ca_3(PO_4)_2$	+4.8 ÷ +6.8

It is seen that these oxides and salts are characterised by weak surface acidic sites. Table 2 shows that thermal treatment yields a change of the acidic strength of some of the fillers.

Table 2. Strength of the acidic sites of thermally treated fillers

Filler	pK_a
$Ca_3(PO_4)_2$ (393 K, 0.1 MPa)	+3.3 ÷ +4.8
$Ca_3(PO_4)_2$ (623 K, 0.1 MPa)	+2.0 ÷ +3.3
γ-Al_2O_3 (393 K, 0.1 MPa)	+2.0 ÷ +3.3
γ-Al_2O_3 (773 K, 0.1 MPa)	-3.0 ÷ +3.3

As is seen, the acidic site strength for a calcium phosphate and alumina increases after a thermal treatment. Such a strength growth is most significant for alumina. Thermal treatment of other fillers that are given in Table 1 does not cause a change of their acidic site strength. We shall further illustrate how the acidic site strength affects polymerization and composite formation.

Acidic and basic site strength can be determined also by using a thermodesorption of gaseous bases and acids. Site strength in this case is characterized by the temperature of thermodesorption maxima [12] or by the desorption activation energy [47]. Some spectroscopic methods make the assessment of the surface site strength through the absorption band shift that is observed within the spectra of adsorbed compounds (sondes) [21, 23]. This technique, together with the methods of coloured indicators, characterize the acid-base site strength in terms of purely non-thermodynamic quantities.

Some adsorption and calorimetric methods from the group of known thermodynamic methods may be also mentioned here. They relate a site strength to adsorption constants, or to adsorption heat capacity of gaseous bases (ammonia or pyridine) and acids (phenol, for instance) [12, 48].

All these methods can be related to the so-called direct methods of determining the surface site strength of solid acids and bases. However, indirect methods are also known, where site strength could be found through the rate constants of certain model heterogeneous catalytic reactions. These are isomerization of hydrocarbons, dehydration of alcohols, esterification of a phthalic acid, etc. [49].

Although different methods and scales are used for the characterization of surface site strength of solid acids and bases, we may generally conclude that mixed oxides are usually stronger acids or bases than simple ones, while the simple oxides possess acidic or basic sites that are stronger than those of salts.

In any case we should note that the classification of solid acids and bases with respect to strength cannot be taken as a basis for a prediction of their activity during different acid-base reactions. Note that strong sites play a major role in some cases, and sites of moderate or even weak strength are decisive in other cases. Tanabe [12] gives a number of such examples. We shall outline only two of them. During the adsorption on moderate acidic sites, for instance, olefins can polymerize to dimers and trimers. At the same time, olefins coke in the presence of strong sites and do not react with weak sites at all. Another characteristic example is the formation of a formalin from dichloromethane. During the accomplishment of this reaction in the presence of aluminosilicate with strong surface acidic sites, different by-products are obtained. At the same time, when formalin interacts with metal sulphates that are weak acids, the reaction develops most of all in the main direction.

A third characteristic of the surface acidity or basicity of solid fillers is the concentration of sites on their surface. Most often, this concentration is determined by using an amine titration, and by employing the same coloured indicators that are used for the determination of acid-base site strength. *n*-Butylamine and benzylamine are usually used as titrants for solid acids, while benzoic and trifluoroacetic acid - for solid bases [12]. Titration is performed from neutral (most often unpolar) solvents. This is done to avoid a possible interaction of the solvent with the titrant and adsorbent. Otherwise, the determined concentration of the acidic or basic sites may be lower than the real one. If we use titrant and solvent of one and the same nature (acidic or basic), it is possible that some of the sites interacts with the solvent and not with the titrant. This is a result of the competing adsorption which occurs. On the other hand, if a titrant nature is opposite to that of the solvent from an acid-base point of view, it is most probable that a part of the titrant interacts with the solvent and not with the adsorbent. Thus, we may obtain unreliable results. Applying the *n*-butylamine titration method, we determined the concentration of the filler surface sites [46]. The results are given in Table 3 For TiO_2 and $Ca_3(PO_4)_2$ the concentration is equal to 0.015 meq/g, while for SiO_2, $BaSO_4$ and $CaSO_4.2H_2O$ - it is by 0.005 meq/g each, respectively. After a filler thermal treatment under conditions, given in Table 2, the concentration for SiO_2 grows to 0.015 meq/g while for γ-Al_2O_3 - from zero to 0.15 meq/g. For all other thermally treated fillers this method does not register a perceptible change in the concentration of filler surface acidic sites.

Note that the approaches of Gutmann and Drago [51, 52] are still used in practice to determine acidic and basic strength. The essence of Gutmann's approach consists in the determination of the donor number (*DN*) of bases and the acceptor number (*AN*) of acids. *DN* is determined by a calorimetric measurements, and *AN* is itself determined by the nuclear magnetic resonance.

Following Gutmann, the heat of a reaction between acids and bases is determined by using the equation

$$-\Delta H_{ab} = AN \times DN / 100 . \tag{7}$$

Table 3. Acidity of inorganic surfaces

Specimen	Concentration of the acidic sites, meq/g, for pK_a			
	-3.0	**+3.3**	**+4.8**	**+6.8**
TiO_2	-	-	0.005	0.010
SiO_2 (quartz)	-	0.005	-	-
γ-Al_2O_3	-	-	-	-
$BaSO_4$	-	-	0.005	-
$CaSO_4 \times 2H_2O$	-	-	0.005	-
$Ca_3(PO_4)_2$ – I sort	-	-	-	0.005
$Ca_3(PO_4)_2$ – II sort	-	-	0.010	0.005
TiO_2, 500 °C	-	-	0.005	0.010
SiO_2, 450 °C	-	0.015	-	-
γ-Al_2O_3, 500 °C	0.080	0.070	-	-
$BaSO_4$, 120 °C	-	-	0.005	-
$CaSO_4 \times 2H_2O$, 120 °C	-	-	0.005	-
$Ca_3(PO_4)_2$ - I sort, 120 °C	-	0.005	-	-
$Ca_3(PO_4)_2$ - II sort, 120 °C	-	-	0.010	0.005

Riddle and Fowkes show in [53] that the contribution to a shift in NMR may be divided into a part, pertaining to dispersion interactions, and one that is due to pure acid-base interactions.

Drago's theoretical approach [51, 52] is based on the linkage between the interaction enthalpy and the acidity and basicity of the interacting surfaces. He introduces two constants (*E* and *C*) in order to determine an acidic or basic strength. These constants are in fact the electrostatic and covalent (i.e. the hard and the soft) contributions to the compound acidity or basicity. Note that the constants are determined here through a calorimetric method. Their linkage with the enthalpy ΔH_{ab} of the reaction between acidic *A* and basic *B* surfaces is given by the equation:

$$-\Delta H_{ab} = C_A C_B + E_A E_B , \tag{8}$$

where C_A and C_B are the acid and base covalent constants, respectively, while E_A and E_B are their corresponding electrostatic constants.

Knowing AN and DN, [22] proposes another empirical method to determine the compound potential at each moment. The method consists in the determination of the parameter K_i:

$$K_i = DN - AN \tag{9}$$

For positive values of K_i, the material displays basic properties, while for negative values of K_i - its properties are acidic.

These three filler characteristics - nature of the surface acidic and/or basic sites, site strength and site concentration, determine filler acidity or basicity. It is essentially noteworthy, however, that each specific case of a filler use is characterized by site nature, strength or concentration, since site display depends also on the acidic or basic properties of the rest of the material components.

3. FILLER ACID-BASE PROPERTIES

In this paragraph we shall analyze the known acidic and basic properties of specific inorganic compounds that are most often used as fillers of polymer composites. The analysis shows that oxides, and particularly silicon dioxide and its different polymorphs, are most often used as fillers for polymers.

It is well known that silicon dioxide exists in three basic allotropic modifications - aerosil, silica gel and quartz. From an acid-base point of view, the presence of isolated, non-interacting hydroxyl groups on the silicon dioxide surface is the common feature of these three modifications [54, 55]. The hydroxyl groups may display proton donating activity that determines the acidic character of the silicon oxides, and the group concentration is up to 7.0 - 9.5 μmol/m^2 or 4.2 - 5.7 groups/nm^2 (100 Å^2). These groups are in fact weak sites of strength between +4.8 and +6.8. Neither Lewis acid sites, nor basic sites are localized on the filler surface. Although the silicon oxides are weak acids, the reaction ability of their hydroxyl groups is strong enough. This ensures a characteristic ability to graft different compounds - modifiers, on the oxide surface. Hence, the prospects of using silicon oxides as effective fillers of polymer composites significantly increase. Among the three polymorphs of the silicon dioxide, aerosil is most commonly used as a polymer filler. This is so, since owing to its large specific surface, the aerosil furnishes the polymer composites with densing and thixotropic abilities, and the polymer processing is relieved as a result. Many of the problems that are related to the surface properties of the silicon oxides are considered in detail in [56-58].

Among other simple oxides that are often used as polymer fillers, we should note aluminium oxide. It exists in amorphous and crystal forms. The γ-crystal modification of the aluminium oxide is most often used for polymer filling. Its introduction into the composite significantly helps to increase polymer hardness and wear resistance. The presence of acidic sites (of protonic and aprotonic type, simultaneously) and basic sites is characteristic for the surface of almost all aluminium oxides. The ratio between them depends on the allotropic modifications of the aluminium oxide, on the method of its obtaining, on the method of oxide surface treatment, on the presence of admixtures, etc. The concentration of protonic acid sites on the surface of various forms of aluminium oxide is smaller than that of the protonic acid site concentration characteristic of silicon oxides. Usually, it does not exceed 1.0 - 1.1 μmol/m^2 or 0.6 - 0.7 groups/nm^2 (100 Å^2). The surface hydroxyl groups are protonic sites. These groups are bonded with the aluminium cations in a different manner, i.e. their coordination state is different. Hence, the result is that the hydroxyl groups are protonized to a different degree and display different acidic strength. As shown in [59], up to 5 types of hydroxyl groups may be identified:

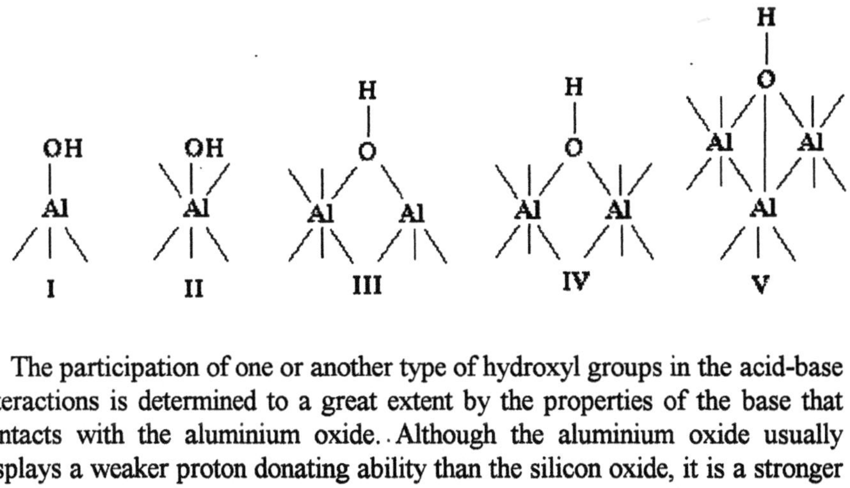

I II III IV V

The participation of one or another type of hydroxyl groups in the acid-base interactions is determined to a great extent by the properties of the base that contacts with the aluminium oxide. Although the aluminium oxide usually displays a weaker proton donating ability than the silicon oxide, it is a stronger acid on the whole. This can be easily explained by the fact that despite protonic sites, strong Lewis acidic sites are located on its surface. These are the coordination unsaturated aluminium cations. Their concentration is up to 2.6 - 5.2 μmol/m^2 or 1.5 - 3.0 sites/100 Å^2. The ratio between the Brønsted and Lewis acidic sites of the aluminium oxide is not a constant quantity, and varies depending on several factors: oxide origin and method of preparation, conditions of surface treatment, etc. For instance, when a thermal treatment of up to 500-1100 K is performed, a dehydroxylation of the surface of the aluminium oxide takes place. Thus, the strength and number of the Lewis acidic sites increase, while those of the Brønsted sites decrease. On the whole, the thermal treatment yields a significant increase of the acidity of the aluminium oxide, since the protonic sites contribute negligibly to the total acidity. The oxide surface hydration produces an opposite effect - the number of Lewis sites decreases and the number of Brønsted sites increases, while the total acidity decreases. The fully hydrated oxide surface does not contain any acidic sites. The aluminium oxide is characterized by its surface bifunctionality. This is due to the location of basic sites on the oxide surface, too - for instance, the negatively charged oxygen atoms of the Al-O-Al group and these of the hydroxyl groups. The concentration is up to 1 - 2 $\mu mol/m^2$ [59].

The structure of the acidic and basic sites that are located on the surface of the aluminium oxide can be schematically illustrated as follows:

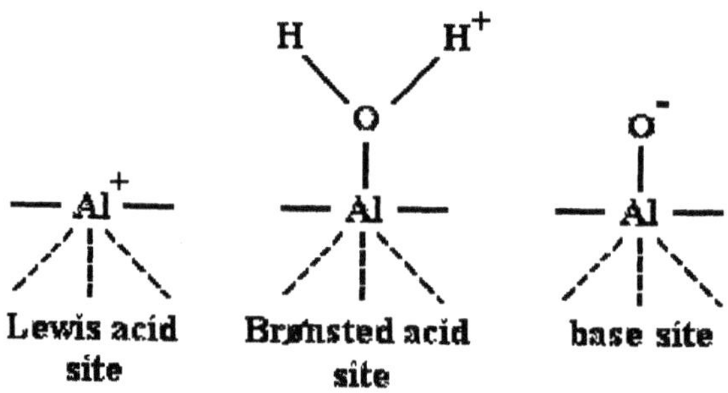

[36-39, 59-66] give a detailed analysis of the various aspects of the surface chemistry of aluminium oxides.

We should note here another dispersed inorganic filler that is often used for polymer filling. This is a titania; it mostly plays the role of a pigment in the polymer compositions. Three crystal modifications of the titania are well known: anatase, rutile and brucite. The forms that are most often used as fillers are the first two. As for aluminium oxide, the surface of the titania may display a bifunctional acid-base character. The presence of acidic and basic hydroxyl groups, basic ions O^{2-} and acidic coordination unsaturated titanium cations are decisive for the compound surface properties. Due to the different crystal structure, the acid-base sites ratio is different for all the crystal modifications of the titania. This ratio significantly changes under an additional thermal treatment. For instance, under definite conditions of thermal treatment, the rutile surface may be formed in such a way that it displays only an acidic character of Lewis type. More details on such type of Lewis acidity are given in the studies of Parfitt, Flaig-Baumann and others [67-72].

A large group of simple oxides is also included in the set of disperse inorganic compounds that are used as fillers for polymer composites. From an acid-base point of view such oxides have been studied much less than silicon, aluminium or titanium oxides.

Hydroxyl groups of an Brønsted acidic character, as well as Lewis acidic sites – coordination unsaturated metal cations, are identified on the surface of ferric and chromium oxides. Besides, weak basic sites (FeOH) are localized in their turn on the surface of the ferrous oxide. Hence, the latter may be considered as a surface amphoteric oxide [73-75]. Interest in the chemistry of ferrous oxide surfaces has strongly increased recently, since it is widely used for the fabrication of magnetic tapes and disks [76].

It is known that magnesium, calcium and zinc oxides may possess only electron-donating ability. This is due to the basic character of the isolated compound surface hydroxyl groups or surface O^{2-}, and no acidic sites have been localized on their surface [77, 78]. These oxides are very effectively used as structurizing fillers for the production of polyester oligomers and rubbers.

Despite the wide spectrum of chemical properties of oxide surfaces, and the variety of ways in which they may be studied, they can be ordered from an acid-base point of view as follows:

- proton donating ability (Brønsted acidity) decreases following the sequence $TiO_2 > SiO_2 > Al_2O_3$;
- electron accepting ability (Lewis acidity) decreases as follows: $Al_2O_3 > TiO_2 > CuO$;
- electron donating ability may be expressed by the following sequence: $CaO > MgO > ZnO > Al_2O_3 \gg TiO_2$.

Mixed or complex oxides are often used as polymer fillers; most popular are aluminosilicates and in particular - clay compounds (kaolinite, bentonite and montmorillonite), since these compounds are easily accessible. They are acidic in character; the origin of their acidic character is complex, and is determined by the location of various exchange cations (K^+, Mg^{2+}, Ca^{2+}, Al^{3+}, etc.) on their surface. A significant number of mobile protons originating from the hydroxyl groups are also present on the compound surface. The exchange cations are generally electron accepting sites, and possess a strong polarizing ability that is characteristic of the aluminium cations. Due to the polarizing activity of the exchange cations, a protolytic dissociation of the molecules of the adsorbed water takes place, and protonic acid sites are formed. Lewis or Brønsted acid sites are present on the surface of clay minerals, depending on the degree of its hydration. Regarding the simple sum of the components composing it (simple oxides), complex oxides display stronger acidity. The presence of weak basic sites on their surface has been established [79, 80].

Another large class of inorganic compounds are various salts that are often used as fillers for polymer composite materials. Sulphates, phosphates, chlorides and carbonates are among them. From an acid-base point of view, their surface properties have been studied less than those of the oxides. As for the phosphates, mostly acid sites of protonic (groups $\equiv$ P-OH) and aprotic types (coordination unsaturated metal cations) are localized on their surfaces. Metal sulphates are usually associated only with the presence of sites of Lewis type, since there are no surface hydroxyl groups. In some cases barium sulphate may display a weak Brønsted acidity due to the presence on its surface of molecules of coordination bonded water [81, 82], while the metal carbonates display only basic properties [83].

Among the fillers for polymers, there are fibers of various types – glass, basalt, carbon, etc. They, like dispersed fillers, may also display acid-base properties.

Various oxides participate in the structure of glass fibers: silicon, aluminium, calcium, magnesium and others. They determine the amphoteric character of their surface. Either acid or basic sites predominate on the glass surface, depending on the type of, and the ratio between, different surface-located oxides. The variety of active surface sites provides for a number of possibilities for glass fiber modification: when fibers are processed with silanes, fiber basicity increases; with a titanate treatment their acidity increases. This in its turn allows for the increase of the adhesion strength between the polymer and fiber, for the increase of the water impermeability of the composite material, and for the improvement of its mechanical characteristics [84-87].

Studies show that the interphase interactions may be strongly affected by using surface-modified organic fibers [88, 89]. Hence, Gaur and Davidson [88] reveal a significant change of the suface properties of aramide fibers after a plasma treatment. Kalantar et.al. [89] demonstarte a significant effect on the properties of aramid fibers after an implantation of Ti^+, Ar^+, N^+ and He^+ ions.

Many attempts have been made to modify the aramid fiber surface, and to promote adhesion, using chemical coupling agents, water boiling, chemical modification, bromine and/or chloric treatment. It is also noted in the literature that the interaction of the aramid fibers with an isocyanate-linked polymer proves to be an effective surface treatment.

Konstantinidou et al. note in [90] that site ratio, type, strength and concentration can be varied during fiber surface treatment within very wide ranges. Separate or combined chemical, radiation and plasma treatments are used for that purpose. Due to some objectives of technical-economic character, the last two methods are still uncommon in practice, although their application is fully legalized. This is so, since aramid plastics possess unique physico-mechanical properties and there are many ways of preparing polymer composite materials with specific properties and application.

A large variety of acidic and basic sites is observed on the surface of carbon fibers that are used as fillers. These are mainly various functional groups - hydroxyl, phenol, carbonyl, carboxyl, etc. [91, 92]. Site ratio, type, strength and concentration may be varied as a result of the additional treatment of the fiber surface, while chemical, radiation and plasma treatment is used for this purpose.

Pigott [93] gives examples of the change of the interphase interactions during the chemical treatment of carbon fibers. He notes that fiber coating can be made of rubbers, soft epoxies, other deformable polymers and viscous layers. All improve fiber toughness but decrease shear strength of the composites. A profound study of the effect of fiber surface treatment is performed by Drzal et al. [94]; he employs XPS analysis, and shows that a chemical adsorption takes place in the presence of both epoxy groups and an amine group. These groups can chemically react with the surface oxygen groups that remain on the surface of the carbon fibers after a fiber oxidation. An interaction that is stronger than the physical bonding is observed during such an interaction. Drzal and co-workers note also that only 4–5 % of the available carbon fiber surface sites can chemically react with the epoxy matrix.

Hanccox notes in [95] that most carbon fibers now undergo an appropriate oxidation treatment of the surface during their fabrication, and are often coated with an epoxy or other polymer-compatible layer to protect the surface and aid bonding. Based on his results [96] and on these of Kardos [97], he supposes that a fiber surface may influence the structure of the surrounding matrix by the development of a preferential adsorption of certain chemicals. Moreover, it acts as a nucleation site for chemical growth, and influences the crystallinity of some thermoplastics.

We should note here that a detailed survey of the influence of carbon, aramid and glass fibers on the nucleation is made by Burton and Folkes [98], considering a variety of polymers (including nylon and polypropylene).

4. ACIDIC AND BASIC PROPERTIES OF MONOMERS, OLIGOMERS AND POLYMERS

These properties of monomers, oligomers and polymers have been studied much less than those of inorganic compounds. Their acidity or basicity has been at most qualitatively characterized; it is mainly the compound chemical structure that has been studied, i.e. the presence of one or another group or links in their molecules that could participate in the acid-base interactions. The acidic and basic properties of monomers and oligomers are estimated on the basis of the compound reaction ability during the catalytic processes of polymerization and polycondensation. The approach proposed by Pearson [99, 100] and Drago [51, 52] has found a limited application. Based on such an approach, acids and bases can be divided into two classes: ***soft*** and ***hard***. Weakly polarizable molecules that have a strong dipole moment may be considered as hard acids and bases, while easily polarizing molecules may be treated as soft acids and bases. Data on the values of parameters *AN* and *DN* enable authors to determine the acid-base properties of some polymers as follows:

- low-density polyethylene is neutral;
- chlorinated polyethylene is acidic;
- polymethylmethacrylate is basic;
- polyvinylchloride is acidic;
- epoxy polymer is amphoteric.

There is another method that is not so popular: the method based on the numerical determination of the values of the electronegativity of separate molecule elements, thus finding the averaged electronegativity of the whole molecule [99, 100]. This mathematical method provides results that strongly differ from the experimental ones. Hence, monomers, oligomers and polymers are usually related either to compounds with basic character, or to compounds with acidic or bifunctional character, depending on the presence of acidic and/or basic functional groups in their molecules. Specific activity in acid-base interactions may display monomers, oligomers and polymers that contain polar groups, such as OH; CN; NH_2; NH; C=O; SH. It is natural, however, that such an activity would depend on the acid-base properties of the introduced filler. Polyolefins, such as polyethylene and polybutadiene, do not contain any substituents and functional groups, and therefore they are not able to participate in acid-base interactions. At the same time, the monomers constituting them may interact with other compounds, following an acid-base mechanism. This is so, since they easily polarize along the double bond in their molecules. When different types of electron-accepting or electron-donating substituents are introduced into the monomer molecules, the compound acidic or basic activity grows significantly.

Regarding other monomers and polymers, we should note an important fact. On the whole, the acidic or basic polymer activity is displayed to an insignificant extent, as compared to that of the corresponding monomers. This is referred mainly to the exhaustion of the functional groups of acidic or basic character in the polymerization process. As a result, the polymer character is determined by the limited number of end groups that are present in the macromolecule and in the monomer molecules that have not reacted yet. Hence, data on monomer acidity or basicity cannot be formally applied to the polymers. This should also be taken into account during the preparation of polymer composite materials, since the filler may be introduced either during the polymerization stage, i.e. in the monomer, or in the ready-made polymer before its further processing. This of course does not concern thermosetting polymers, since the filler in that case can be introduced only in the oligomer.

Regarding the chemical structure and composition of the compounds and generalizing data from literature, it is not hard to establish that monomers, oligomers and polymers are mostly compounds with basic properties; examples include polystyrene, polyacrylates, polycarbonate, polysiloxanes, polytetrafluoroethylene, polyacrylonitrile, polyvynilacetate, polyamides, aminooligomers. The group of monomers, oligomers and polymers that have acidic character is significantly smaller: principally polyvinylchloride and phenolic oligomers. Some of the monomers, oligomers and polymers have a bifunctional character, i.e. they may display acidic, as well as basic properties: polyurethanes, epoxyoligomers and unsaturated polyester oligomers belong to this group.

Polystyrene may display basic character, owing to the phenyl groups that participate in its structure. Polycarbonate contains basic-in-character carbonate groups. Polyacrylates and polyvinylacetate display basic properties due to the contents of carbonyl groups. Amines, as well as carbonyl groups, are responsible for the basicity of the polyamides and amine-containing oligomers. For silicon containing monomers and polymers, lateral substituents in their molecules (phenyl, nitrile and other groups) determine the compound basicity. Polyvinylchloride acidity is due to C-Cl groups, while that of phenolic oligomers is due to phenol, methylol and hydroxyl groups. The bifunctional character of the epoxy oligomers arises from the basic-in-character epoxy groups and the acidic primary and secondary hydroxyl groups that are formed when opening the epoxy ring. Polyurethanes may display, on the one hand, acidic properties due to the presence of hydroxyl and primary isocianate groups in their composition, and on the other hand basic properties that result from the carbonyl and amine groups.

One of the methods for an effective monitoring and control of the interphase interactions at polymer surfaces is an appropriate surface modification. The goal of controlling the surface properties of the polymers has been persued for many years and various methods have been developed for this purpose. Schreiber [22] classified these methods as *mechanical, thermal, physical, chemical and inherent*.

The methods of **mechanical** surface modification comprise simple wash, abrasion treatment of the polymer surfaces, and thermal conditioning procedures. Polymer surfaces that are similar to those of ceramic materials and metals, are in most cases contaminated by various compounds. Generally, dust and chemical contaminating agents are removed by a simple washing, using solvents or non-solvent fluids. Drying procedure follows washing and the solvent liquid is removed from the surface. Solvent wash aims at the removal of asperities from the polymer surface. Hence, conditions for a close contact between the polymer surface and the contacting filler particles are created. These are in fact conditions for a fuller realization of the interphase interactions.

Another method for the removal of loosely bonded contaminants and surface irregularities is a simple abrasion. This procedure should be followed by washing, since the abrasive agent can itself be a contaminant.

Data from literature show that **thermal** conditioning may significantly affect the properties of the polymer surface. Here tempering of the so-called hygroscopic polymers, such as polyesters and polyamides, is necessary. All these should be dried before using them in composites, in order to remove (partially or completely) the physically or chemically sorbed water. In the literature, special attention has been paid to the thermal conditioning of polyolefins, the density of which is notably low, and to the treatment of a linear low-density polyethylene [22].

The **physical** modification comprises two approaches (i.e. two procedures): glow discharges at atmospheric, and reduced pressure. The most often used procedure for the conditioning of the polymer surface is that in which discharges are operated at atmospheric pressure. This method is known in the literature as a corona discharge. Such a treatment is linked to a transition of the polymer solid through a gap, the dimensions of which are usually millimetres. Such a gap exists between earthen and loaded electrodes. The electrodes generally have a potential of 15 kV, and generate current of about 10 - 50 μA. The atmospheric gases (oxygen and nitrogen) under these conditions ionize and interact with the modified substrate. As the experiments show, the exposure time necessary for the surface modification, is within the range 1 - 30 s; it is related to the chemical processes that may occur under corona discharges, but also depends on the type of the modified substrate. As is shown in the literature, this method is most often applied for the treatment of polyolefin surfaces on purpose to formate acid-base surface sites [22].

As is known, linear low-density polyethylene (LLDPE) has neither acidic nor basic interaction tendencies, and its surface energy is due only to the dispersion force. Corona treated polyolefin can interact with polar wetting fluids after the treatment execution. As Schreiber notes in [22], corona treatment may play a decisive role in the formation of polymer composites. The author gives an example of the successive application of cellulose fibers as a reinforcing constituent of polymer composites. These fibers are characterized by acidic sites. Their treatment with corona discharges, following Uehara [101], may significantly modify their surface energy. The result is a formation of composites with the polyolefins, characterized by significantly higher mechanical properties [22].

Discharge operation at a reduced pressure at ratio -16.56 MHz or microwave - 2.45 GHz frequency yields an important surface modification. This method is also known as a cold plasma discharge. It is tackled in detail in Boenig's monograph [102] and is used for surface cleansing ablation, surface modification and plasma deposition or polymerization.

Chemical modification is most often used when coupling or compatibilizing agents are applied. These compounds comprise a wide set of chemicals: titanates, zircoaluminates, as well as similar bi- or multifunctional molecules. The best known set, however, is the large group of silanes. Their structure and ability to modify solid surfaces by acting as bridging or coupling agents that bond dissimilar materials are described in detail in [103, 104]. At the same time, it is noted in literature [22], that a possible mechanism of their action is the acid-base interaction. Simultaneously, the author notes that the problem of elucidation of the mechanism of its operation is too complex, and there are various contradictory ideas. It is also considered that the most important mechanism appears to be the formation of covalent bonds between the silane and the modified surface of polymers, glass, minerals, etc.

We should also note that the formation of covalent bonds is not a necessary condition. It is possible that the interaction between glass fibers and some thermoplastics can be strongly promoted by silane agents, even though no evidence for covalent bonding can be found [103].

The silanes considered in [22] are classified as *monomeric* compounds. A better effect may be guaranteed during the improvement of the adhesion of joints that are difficult to bond by using polymeric silanes, i.e. Si-O linkages associated with polymers such as polyethyleneimine. Thus, strong and stable bonds are formed between materials that normally may not bond adequately. Parallel to this advantage of *silane* bond improvement, we should note the limited ability of the material to withstand long-term temperature effects under a temperature higher than 250 ^{O}C. An increase of the material temperature resistance may be achieved by using the advantages of high-performance highly temperature-resistant polymers. As is known, polyamides, polyethers and polyketones belong to such a group.

The so-called **inherent** modifications have good prospects for polymer modification. They are related to the ability of a number of polymers to change their surface properties, while polymer surface restructuring takes place under the action of a contacting medium. Such surface modifications are based on the thermodynamics of interfaces and interphase interactions within multicomponent polymers. In [22, 105] is noted that the acid-base interactions may yield surface (or inherent) transformation of the polymer molecule. This is accomplished at the expense of the selective interaction between some solvents (water, formaldehyde, etc.) and specific polymer fractions; mainly it is employed in two-component polymers, such as polyurethanes. As a result of such interactions a change of the polymer surface energy occurs, the polymer wetting ability increases and the adhesion strength grows.

Note that this modification of polymer surface properties is not directly related to the problems we tackle in our book, but is of a great importance for the formation of polymer coatings (and for filled polymeric coatings, also). Moreover, such a modification suggests additional possibilities for the control of polymer composite formation and enables one to use an acid-base approach when selecting the main polymer components.

REFERENCES

1. *Interface in Polymer Matrix Composites*. E. P. Plueddmann, ed. N. Y.-London: Academic press, 1974.
2. Manson, John, Sperling, Leslie, *Polymer Blends and Composites*. N.Y., London: Plenum Press, 1976.
3. Jensen, William, *The Lewis Acid-Base Concepts: An Overview*. N. Y.: Wiley-Interscience, 1980.
4. *Molecular Characterization of Composite Interfaces*. H. Ishida and G. Kumar, eds. N. Y.: Plenum, 1985.
5. *Polymer Composites*. B. Sedlacek, ed. Berlin-N. Y.: W. de Gruyter, 1986.
6. Briscoe, Brian, Roach, S., Williams, D. R. *Interfacial Phenomena in Composite Materials*. Guilford: Butterworth, 1989.
7. *Acid-Base Interactions: Relevance to Adhesion Science and Technology*. K. L. Mittal and H. R. Anderson, eds. Utrecht: VSP, 1991.
8. *The Interfacial Interactions in Polymeric Composites*. G. Akovali, ed. NATO Advanced Study Institute, Series E: Applied Sciences, 1993.
9. *Particulate Filled Polymer Composites*. R. N. Rothon, ed. Longman, 1995.
10. Fowkes F.M. Acid-base contributions to polymer-filler interactions. Rubber Chem. and Technol. 1984; **57**: 328-42.
11. Manson J.A. Interfacial effects in composites. Pure and Appl. Chem. 1985; **57**: 1667-78.
12. Tanabe, Kozo, *Solid Acids and Bases. Their Catalytic Properties*. N. Y.: Academic Press, 1970.
13. Bakshi K.R., Galavas G.R. Characterization of acid-base catalysts by calorimetric titration. 1. Correlation with alcohol dehydratation activity. J. Catal. 1976; **38**: 312-25.
14. Kijenski J., Baiker A. Acidic sites on catalysts surface and their determination. Catalysis Today 1989; **5**.
15. Benesi H.A., Winquist B.H.C.: Surface acidity of solid catalysts. Adv. Catal. 1978; **27**: 97-182.
16. Krzywicki A., Marczewski M., Modzelewski R., Pelszik K., Malinowski R. Superacid alumina catalysts. React. Kinet. Catal. Lett. 1980; **13**: 1-6.
17. Krzywicki A., Marczewski M. Superacidity of modified γ-alumina. Structure of active site and catalytic activity. J. Chem. Soc. Faraday Transl. 1 1980; **76**: 1311-22.
18. Jacobs, Peter. *The Measurement of Surface Catalysts*. N. Y.: Marcell Dekker , 1984.
19. Metulewics E.R.A., Kerkhof F.P.J.M., Moulijn J.A., Reitsma H.J. Structure and activity of fluorinated alumina. 1. Determination of number of protonic sites by an infrared study of adsorbed pyridines. J. Colloid Interfaces 1980; **77**: 110-19.
20. Pleifer H., Hunger M., Freude D. Nuclear magnetic resonance studies on the acidity of zeolites and related catalysts. Zeolites 1985; **5**: 274-86.
21. Mastihin V.M., Mudrakovski I.L., Filimonova S.V. Probing the Lewis acidity of heterogeneous catalysts by nitrogen-15 NMR of adsorbed nitrous oxide. Chem. Phys. Lett. 1988; **149**: 175-79.
22. Schreiber, H. P. "Aspects of Component Interactions in Polymer Systems." In: *Interfacial Interaction in Polymeric Composites*. G. Akovali, ed. Netherlands: Kluwer Acad. Publ., 1993, 21-59.
23. Angell C.L., Howell M.V. Infrared spectroscopic investigation of zeolites and adsorbed molecules. IV. Acetonitrile. J. Phys. Chem. 1969; **73**: 2551-54.
24. Peri J.B. The effect of fluoride on surface "acid" sites on γ-alumina and silica-alumina. J. Phys. Chem. 1968; **72**: 2917-25.
25. Peri J.B. A model for the surface of γ-alumina. J. Phys. Chem. 1965; **69**: 220-30.

26. Hair M., Hertle W. Acidity of surface hydroxyl groups. J. Phys. Chem. 1970; **74**: 91-94.
27. Paukstis E.A., Karakchiev L.G., Kotsarenko N.S. Proton donor ability of surface hydroxyl groups from the IR spectra of hydrogen-bonded complexes. React. Kinet. Catal. Lett. 1978; **9**: 265-68.
28. Burgina E.B., Yurchenko E.N., Paukstis E.A. Peculiarities of the vibrational spectra of molecules adsorbed on the surface of heterogeneous catalysts: theoretical analysis. J. Mol. Structure 1986; **147**: 193-201.
29. Uytterhoeven J.B., Christner L.G., Hall W.K. Studies of the hydrogen held by solids. VIII. The decationated zeolites. J. Phys. Chem. 1965; **69**: 2117-26.
30. Basila M.R., Kantner T.R., Rhee K.H. The nature of the acidic sites on a silica-alumina. Characterization by infrared spectroscopic studies of trimethylamine and pyridine chemisorption. J. Phys. Chem. 1964; **68**: 3197-207.
31. Hughes T.R., White H.M. A study of the surface structure of decationized Y zeolite by quantitative infrared spectroscopy. J. Phys. Chem. 1967; **71**: 2192-201.
32. Jacobs P.A., Uytterhoeven J.B. Quantative infrared spectroscopy of amines in synthetic zeolites X and Y. I. Alkylammonium Y zeolites as precursors of acid hydroxyls in deaminated zeolites Y. J. Catal. 1972; **26**: 175-90.
33. Jacobs P.A., Theng B.K.G., Uytterhoeven J.B. Quantitative infrared spectroscopy of amines in synthetic zeolites X and Y. J. Catal. 1972; **26**: 191-201.
34. Purcell K.F., Drago R.S. Studies of the bonding in acetonitrile adducts. J. Am. Chem. Soc. 1966; **88**: 919-29.
35. Zenkovets G.A., Paukstis E., Tarasova D.V., Nikoro T.A., Khomicheva S.Ya. Acid-base and catalytic properties of iron-antimony oxide catalysts. React. Kinet. Catal. Lett. 1981; **16**: 142-46.
36. Zaki M.I., Knoesinger H. Carbon monoxide – a low temperature infrared probe for characterization of hydroxyl group properties on metal oxide surfaces. Mater. Chem. and Phys. 1987; **17**: 201-15.
37. Della Gatta G., Fubini B., Chiotti C., Mortera C. The chemisorption of carbon monoxide on various transition aluminas. J. Catal. 1976; **43**: 90-7.
38. Scokart P.O., Rouxhet R.G. Characterization of the basity of oxides through the infrared study of pyrrole adsorption. J. Chem. Soc. Faraday Trans. 1 1980; **76**: 1476-89.
39. Paukstis E.A., Kotsarenko N.S., Karakchiev L.G. Investigation of proton-acceptor properties of oxide surfaces by IR spectroscopy of hydrogen-bonded complexes. React. Kinet. Catal. Lett. 1979; **12**: 315-19.
40. Tanabe, Kozo. "Solid Acid and Base Catalysis - Science and Technology." In *Catalysis: Science and Technology*. **2**. J. R. Anderson and M. Bondart, eds. Berlin: Academic Verlag, 1983.
41. Gervasini A., Auroux A. Microcalorimetric investigation of the acidity and basicity of metal oxides. J. Therm. Anal. 1991; **37**: 1737-44.
42. Kustov L.M. New trends in IR-spectroscopic studies of acidic and basic centers in zeolitic and oxide catalysts. Russ. Chem. J. 1998; **1-2**: 163-75 (In Russian).
43. Ismail H.M., Mansour S.A.A., Zaki M.J. Acid properties of silica and alumina surfaces as probed by thermogravimetry and DSC. Thermochim. Acta 1992; **202**: 269-80.
44. Walling C. The acid strength of surfaces. J. Am. Chem. Soc. 1950; **72**: 1164-68.
45. Deeba M., Hall W.K. The measurement of catalyst acidity. 1. Titration measurement. J. Catal. 1979; **60**: 417-29.
46. Cheshkov VM, Natova MM. Acid-base interactions in filled polymers. Proceedings of the VII International Conference on Mechanics and Technology of Composite Materials; 1994 October 10 - 12; Sofia: CLPhChM, 1994.
47. Kubokawa Y. Determination of acidity of solid catalysts by ammonia chemisorption. J. Phys. Chem. 1963; **67**: 769-71.

48. Hsieh P. Y. Heats of adsorption of ammonia on silica-alumina catalysts and their surface energy distributions. J. Catal. 1963; **2**: 211-22.
49. Forni L. Comparison of methods for the determination of surface acidity of solid catalysts. Catal. Rev. Sci. Eng. 1973; **8**: 65-115.
50. Gutmann V., Steininger A., Wychera E. Donor strengths in 1,2-dichloroethane. Monatsh. Chem. 1966; 97: 460-67.
51. Drago R.S., Vogel G.C., Needham T.E. A four-parameters equation for predicting enthalpies of adduct formation. J. Am. Chem. Soc. 1971; **93**: 6014-26.
52. Drago R.S., Parr L.B., Chamberlain C.S. Solvent effects and their relationship to the E and C equation. J. Am. Chem. Soc. 1977; **99**: 3203-9.
53. Riddle F. L., Fowkes F. M. Spectral shifts in acid-base chemistry. 1. van der Waals contributions to acceptor number. J. Am. Chem. Soc. 1990; 112: 3259-64.
54. Garrone E., Kazanskii V.B., Kustov L.M., Sauer J., Sencheya L.N., Ugliengo P. Spectroscopic and *ab initio* study of the interaction of molecular hydrogen with the isolated silica hydroxyls and related systems. J. Phys. Chem. 1992; **96**: 1040-45.
55. Morrow B.A., Cody I.A. Infrared spectra of the isolated hydroxyl groups on silica. J. Phys. Chem. 1973; **77**:1465-69.
56. Paukstis E.A., Yurchenko E.N. Estimation of proton donor ability of surface hydroxyl groups in terms of proton affinity. React. Kinet. Catal. Lett. 1981; **16**: 131-35.
57. Ligner G., Vidal A., Ballard H., Papirer E. London component of surface energy of heat-treated silicas. J. Colloid Interfaces 1980; **133**: 200-10.
58. Papirer E., Balard H. "Influence of Surface Chemistry and Surface Morphology on the Acid-Base Interaction Capacities of Glass Fibers and Silicas." In *Acid-Base Interactions: Relevance to Adhesion Science and Technology*. K. L. Mittal and H. R. Anderson, eds., Utrecht: VSP BV, 1991.
59. Knösinger H., Ratnasamy O. Catalytic aluminas: surface models and characterization of surface sites. Catal. Rev. -Sci. Ing. 1978; **17**: 31-70.
60. Fleisher M.B., Golender L.O., Shimanskaja M.V. Quantum chemical models for basic sites on the surface of γ-alumina catalysts. React. Kinet. Catal. Lett. 1987; **34**: 137-42.
61. Paukstis E.A., Soltanov R.I., Yurchenko E.N. Determination of the strength of aprotic acidic centers on catalyst surfaces from the IR spectra of adsorbed carbon monoxide. React. Kinet. Catal. Lett. 1981; **16**: 93-5.
62. Paukstis E.A., Soltanov R.L., Yurchenko E.N. Determination of the strength of Lewis acid centers via IR spectroscopic measurement of adsorbed pyridine. React. Kinet. Catal. Lett. 1982; **19**: 105-8.
63. Paukstis E.A., Soltanov R.L., Yurchenko E.N. IR spectroscopic determination of protonation heats of bases on catalyst surfaces. React. Kinet. Catal. Lett. 1982; **19**: 119-23.
64. Kawakami H., Yoshida S. Quantum-chemical studies of alumina. Part 2. Lewis acidity. J. Chem. Soc. Faraday Trans. 1985; **81**: 1129-37.
65. Kawakami H., Yoshida S. Quantum-chemical studies of alumina. Part 1. Broensted acidity and basicity. J. Chem. Soc., Faraday Trans. 1985; **81**: 1117-27.
66. Vit Z., Vala J., Malek J. Acid-base properties of aluminium oxide. Appl. Catal. 1983; **7**: 159-68.
67. Parfitt G.D., Ramsbotham J., Rochester C.H. An infrared study of pyridine adsorption on rutile surface. Trans. Faraday Soc. 1971; **67**: 1500-06.
68. Flaig-Baumann R., Neuwinger H.D., Boehm H.P. Amphotere eigenschaften von metalloxid-oberflächen. Fortschr. Kolloide und Polym. 1971; **B55**: 7-15.
69. Jones P., Hockey J. Infrared studies of rutile surfaces. Pt. 2. Hydroxylation, hydratation and structure of rutile surface. Trans. Faraday Soc. 1971; **67**: 2679-85.

70. Donnet J.B., Muller F. Properties physico-chimique superficielles des oxides de titane. II. Détermination des sites actifs á la surface des cristaux d'oxyde de titane pigmentaire. Bull. Soc. Chim. France, Pt. 1 1973; **9/10**: 2605-10.
71. Scokart P.O., Rouxhet P.G. Comparison of the acid-base properties of various oxides and chemically treated oxides. J. Colloid and Interface 1982; **86**: 96-104.
72. Mason J.G., Suriwardane R., Wightman J.P. Spectroscopic characterization of acidity of titanium (6 % aluminium – 4 % vanadium) surfaces. J. Adhesion 1981; **11**: 315-28.
73. Micale F.J., Kiernan D., Zettlemoyer A.C. Characterization of surface properties of iron oxides. J. Colloid and Interface Sci. 1985; **105**: 570-76.
74. Zecchina A., Coluccia S., Guglielminotti E., Ghiotti G. An infrared study of surface properties of α-chromia. I. Preparation and adsorption of water, heavy water and carbon monoxide. J. Phys. Chem. 1971; **75**: 2774-83.
75. Zecchina A., Guglielminotti E., Gerrutti L., Coluccia S. An infrared study of surface properties of α-chromia. IV. Pyridine and heavy water adsorption on oxygen-covered surface. J. Phys. Chem. 1972; **76**: 571-77.
76. Huang, Y. C., Fowkes, F. M., Lloid, T. B. "Acid and Base Nature of Ferric Oxide Surfaces. Adsorption, Adhesion, Zeta Potentials and Dispersibility in Magnetic Inks for Hard Disks." In *Acid-Base Interactions: Relevance to Adhesion Science and Technology*. K. L. Mittal and H. R. Anderson, Jr., eds. Utrecht: VSP, 1991.
77. Morishige K., Kittaka S., Moriyasu T. Thermal desorption study of surface hydroxyls on zink oxide. J. Chem. Soc. Faraday Trans. 1 1980; **76**: 728-45.
78. Che M., Naccahe C., Imelik B. Spin resonance studies a titanium dioxide and magnesium oxide-electron donor properties. J. Catal. 1972; **24**: 328-35.
79. Mizuno K., Ikeda M., Imokawa T., Take J., Yoneda Y. Nature of catalytically active sites over solid acids. I. Selective poisoning of Lewis acid sites on silica-alumina with pyridine and its application to olefin polymerization. Bull. Chem. Soc. Jap. 1976; **49**: 1788-93.
80. Take J., Ikeda M., Yoneda Y. Nature of catalytically active sites over solid acids. II. Relationships between acidic properties of silica-alumina and its catalytic activities for olefin polymerization. Bull. Chem. Soc. Jap. 1977; **50**: 49-51.
81. Negle W., Rochester C.H. Infrared and gravimetric studies of the adsorption of water on barium sulphate. J. Chem. Soc. Faraday Trans. 1 1988; **84**: 3625-32.
82. Negle W., Rochester C.H. Infrared study of the adsorption of ammonia, pyridine and hydrogen chloride on barium sulphate. J. Chem. Soc. Faraday Trans. 1 1988; **84**: 3615-23.
83. Schreiber H.P., German F.St. "Specific Interactions and Their Effect on the Properties of Filled Polymers." In *Acid-Base Interactions: Relevance to Adhesion Science and Technology*. K. L. Mittal and H. R. Anderson, Jr., eds. Utrecht: VSP, 1991.
84. Fowkes F.M., Dwight D.W., Cole D.A., Huang T.C. Acid-base properties of glass surfaces. J. Non-Cryst. Solids 1990; **120**: 47-60.
85. Osmont E., Schreiber H.P. "Surface characteristics of glass fibers." In *Inverse Gas Chromatography*, D. R. Lloid, T. C. Ward and H. P. Schreiber, eds. ACS Symposium Series 391, 230-247, Am. Chem. Soc. Washington, DC, 1989.
86. Saint Flour C., Papirer E. Gas-solid chromatography: a quick method of estimating surface free energy variations induced by the treatment of short glass fibers. J. Colloid Int. Sci. 1983; **91**: 69-75.
87. Fowkes F.M., McCarthy D.C., Tischer D.O. "Predicting entalpies of interfacial bonding of polymers to reinforcing pigments." In *Molecular Characterization of Composite Interfaces*, H. Ishida and G. Kumar, eds. N. Y.: Plenum Press, 1985.
88. Davidson G.T. "Interfacial Effects of Plasma Treatment of Fiber Pull-out." In *Interfaces in Composites*. C. Pantano and E. Chen, eds. Pittsburgh: MRS, 1989.

89. Kalantar J., Grummon D., Drzal L., Loh I., Moody R. "Effect of Ti^+, Ar^+, N^+ and He^+ Ion Implantation on Aramid Fiber Adhesive Properties." In *Interfaces in Composites*. C. Pantano and E. Chen (eds.) MRS, Pittsburgh, 1989, 315-327.
90. Konstantinidou A.V., Petsalas H.J., Andreopoulos A.G. "Kevlar - Polyurethane Composites: a New Incorporation Technique." In *Phase Interaction in Composite Materials*. S. A. Paipetis and G. C. Papanicolau, eds. England: Omega Scientific, 1992.
91. Donnet, Jean-Baptiste, Bansal, Roup. "International Fiber Science and Technology." In *Carbon Fibers*, **3**. N. Y.: Marcell Dekker, 1984.
92. Papirer E., Li S., Donnet J.-B. Contribution to the study of basic surface groups on carbons. Carbon 1987; **25**: 243-47.
93. Pigott M. Tailored Interfaces in Fiber Reinforced Polymers, in: *Interfaces in Composites*. C. Pantano and E. Chen, eds. Pittsburgh: MRS, 1989.
94. Drzal L.T., Rich M.J., Lloyd P.E. Adhesion of graphite fibres to epoxy matrices. I. The role of the surfaces treatment. J. of Adhesion 1982; **16**: 1-30.
95. Hanccox, N.L. "The Influence of Fiber-Matrix-Interaction in Determining Mechanical Properties of Composites." In *Phase Interaction in Composite Materials*. S. A. Paipetis and G. C. Papanicolaou, eds. England: Omega Scientific, 1992.
96. Hanccox N.L. The effect of flaws and voids on the shear properties of CFRP. J. of Material Science 1977; **12**: 884-92.
97. Kardos, J.L. "The Role of the Interface in Polymer Composites - Some Myths, Mechanisms and Modifications." In *Molecular Characterization of Composite Interfaces*. Eds. H. Ishida and C. Kumar, eds. N. Y.: Plenum, 1985.
98. Burton, R., Folkes, M. "Fiber-Matrix Interaction in Reinforced Thermoplastics." In *Mechanical Properties of Reinforced Thermoplastics*. D. Clegg and A. Collyer, eds. London: Elsevier Appl. Sci., 1986.
99. Pearson R.G. Hard and soft acids and bases (HSAB). I. Fundamental principles. J. Chem. Educ. 1968; **45**: 581-87.
100. Pearson R.G. Hard and soft acids and bases (HSAB). II. Underlying theories. J. Chem. Educ. 1968; **45**: 643-48.
101. Uehara T., Sakata I. Effect of corona discharge treatment on cellulose fibers prepared from beechwood. J. Appl. Polym. Sci. 1990; **41**: 1695-1706.
102. Boenig, Herman, *Advance in Low-Temperature Plasma Chemistry, Technology, Applications*. Lancaster: Technomic Publishing Co., 1984.
103. Plueddemann, Edwin, *Silane Coupling Agents*. N. Y.: Plenum Press, 1990.
104. Plueddemann, Edwin, *Composite Materials* **6**. N. Y.: Academic Press, 1974.
105. Chen P.J.H., Ruckenstein E. Solvent-stimulated surface rearrangement of polyurethanes. J. Coll. Int. Sci. 1990; **135**: 496-507.

Chapter 2

ADSORPTION ACID-BASE INTERACTIONS IN FILLED COMPOSITES

The behaviour of composite materials depends on the nature of the interaction between the separate components during the composite system formation. There are many contradictory results; often, different researchers cannot repeat them. What is determining during the system formation is the acid-base character of the interaction between the separate compounds. Such an interaction develops not only within each phase of the system, but also on the interfaces. Such a concept must be applied to polymer formation in the presence of a solid surface; the latter, being one of the process agents (a solid acid or base), should participate, too, in the surface interactions (acidic or basic ones). The reason that many results often cannot be confirmed or predicted could be explained by establishing how the acid-base interactions affect both the filling progress and the whole polymer system.

A significant factor that affects the interaction between monomers, oligomers and polymers, on one hand, and the solid, most often inorganic admixtures (fillers), on the other hand, is the formation of links of different character in the system (e.g. during sorption). Link formation is determined by various acidic or basic sites located on the inorganic surfaces. The existence of molecules adsorbed on the solid surface yields a redistribution of the electron density of the compounds and finally - formation of more *inclusion reaction* compound modes. Thus, the adsorption phenomena play a major role in the synthesis of polymer materials that takes place in the presence of solid substances - inorganic fillers.

1. THEORETICAL BASE OF ADSORPTION ON SOLIDS

The studies of adsorption from a solution and onto the surface of a solid substance can disclose essential characteristics of the interactions that take place in such heterogeneous *liquid-solid* systems. Intermolecular forces that act on the interface develop an adsorption process on the surface. During the adsorption process, molecules of the adsorbate leave the gas phase, solution or liquid phase, approach the interface driven by forces of attraction, and settle there. Thus, the surface of the adsorbent (the solid phase) is covered by a thin (adsorption) film of molecules of the liquid (adsorbate); attractive forces prevail over repulsive forces during the molecule drive to the interface. The adsorption interactions represent one of the most important stages of the heterogeneous catalysis. Hence, it is important to study the kinetics of these phenomena and the changes that occur when varying adsorption conditions.

Change of the temperature affects significantly adsorption. One can calculate the thermal effects and estimate the energy interactions within the system, using the adsorption temperature dependence. To calculate the isosteric temperature of adsorption, one may use the Clausius-Clapeyron equation

$$\frac{\Delta H}{R} = \frac{d \ln \frac{a_1}{a_0}}{d\left(\frac{1}{T}\right)} \qquad (2.1)$$

where ΔH is the difference between the partial mole enthalpies on the adsorbent surface and within the solution, while a_1 and a_0 are, respectively, the substance activities within the liquid and on the interface.

Extract now the concentrations, corresponding to equal quantities of the adsorbed substance, from the adsorption isotherms obtained for different temperatures. Plot them on the graph $c = f(1/T)$. Then, the slope of the obtained straight line will produce the enthalpy change occurring during the "solution volume - surface" molecule transfer. The slope also characterizes the overall energy of process activation:

$$E = -2.3\,R\left|tg\,\alpha\right|, \qquad (2.2)$$

where R is the universal gas constant.

Due to the similarity between the shape of the gas phase isotherms and those of diluted solutions, the latter can be mathematically modelled by employing Langmuir's equation for monomolecular adsorption. This equation is derived under a number of assumptions, of which the most important are as follows:

a) the adsorbent surface is energetically homogeneous, i.e. it is a set of energetically equal active sites where the adsorbate molecules adsorb;
b) a monolayer develops during adsorption, only, i.e. molecules adsorb on sites located on the free surface, and they are elastically repulsed back into the solution by the adsorbate-occupied areas;
c) the interaction between adsorbed molecules is disregarded, i.e. adsorption on a given site does not affect the adsorption on other sites.

If the equilibrium during adsorption from a diluted solution can be expressed by the scheme

$$\begin{array}{ccccc} \text{Molecule} & & \text{Free adsorption} & & \text{Adsorption} \\ & + & & \leftrightarrow & \\ \text{in a solution} & & \text{site} & & \text{complex} \end{array} \qquad (2.3)$$

i.e. if a dynamic equilibrium between adsorption and desorption is accomplished, then the constant of adsorption equilibrium K_p can be expressed in terms of the molecule concentration in the solution c, by the concentration of free sites $(A_m - A)$ and the surface concentration of occupied adsorption sites A:

$$K_p = \frac{c\,(A_m - A)}{A}, \qquad (2.4)$$

where A_m is the total number of adsorption sites. From this relation we can get the general Langmuir equation

$$A = A_m \frac{c.b}{1 + c.b}, \qquad (2.5)$$

or

$$c = \frac{A}{(A_m - A).b}, \qquad (2.6)$$

where $b = K_p^{-1}$ is so called adsorption coefficient.

For very low solution concentrations, the value $(c.b) \to 0$, $c.b << 1$ and $(1 + c.b) \to 1$, respectively, and then

$$A = (A_m . b)$$

or

$$A \to A_m, \tag{2.7}$$

Relation (2.7) gives a linear dependence between the adsorption and the concentration. For Langmuir's equation, adsorption is equal to half of the maximal value, for $c = 1/b$

$$A = \frac{A_m}{2} \tag{2.8}$$

For high solution concentrations (where interaction phenomena are closer in character to the limit case of interaction between solid substances) $c >> 1/b$, i.e. $c.b >> 1$. Then, the unit in the denominator of equation (2.5) can be disregarded and we get, that

$$A \approx A_m, \tag{2.9}$$

i.e. the adsorption does not depend on the solution concentration.

To determine the constants A_m and b that are included in the Langmuir equation, using the experimental dependence for $A = f(c)$, one can rewrite equation (2.4) in the form:

$$\frac{c}{A} = \frac{1}{A_m b} + \frac{1}{A_m}.c \tag{2.10}$$

or

$$\frac{1}{A} = \frac{1}{A_m .b.c} + \frac{1}{A_m}. \tag{2.11}$$

Hence, the experimental data form a straight line with coordinates $1/A = f(1/c)$. The value of A_m represents a ordinate section cut by this line, while b is the tangent of the line slope angle.

The mathematical processing of the experimental isotherms is employed mainly to determine the monolayer capacity, i.e. the value of A_m corresponding to the quantity of the substance filling the first layer, i.e. the layer that is immediately next to the adsorbent surface. The value of A_m, however, gives an approximate and most general assessment of the accessible adsorbent surface.

Equation (2.5) corresponds strictly to the adsorption model proposed by Langmuir. Owing to the process idealization, it can seldom describe the experimentally obtained results with a sufficient accuracy. To describe the adsorption on a non-homogeneous surface, i.e. under the assumption that not all the adsorption sites are identical, one has to use the theory of polymolecular adsorption, proposed by S. Brunauer, P. Emmett and A. Taylor [1, 2]. Following this theory, each molecule, adsorbed in the first layer, becomes a site for molecules that are to form the second layer and so on. This model is mathematically set forth by the equation:

$$\frac{p}{v(p_0 - p)} = \frac{1}{v_m \cdot c} + \frac{(c - 1)}{v_m \cdot c} \cdot \frac{p}{p_0}, \qquad (2.12)$$

where v is the volume of the adsorbed substance, v_m is the capacity of the monolayer, p is the equilibrium pressure and c is a constant.

This gives a straight line representation as $y = mx + a$, where $y = p/p_0 - p$,

$$x = p/p_0, \quad m = \frac{c-1}{v_m \cdot c}, \quad a = \frac{1}{v_m \cdot c}.$$

2. ADSORPTION INTERACTIONS IN FILLED POLYMER SYSTEMS

It has long been known that an adsorbate is a special state of matter, different from the normal state of the volume phases [3-15]. Having generalized and analyzed the experience gained in the field, we distinguish two basic approaches to clarifying and setting forth the details of the interaction processes that take place in filled polymer systems. The first one comprises studies of the adsorption of previously prepared polymers onto inorganic fillers, and explains the reasons for the resulting changes of the system. The second approach yields, first of all, a separate study of adsorption. Then, it involves an analysis of adsorption interactions that occur during the formation of systems of filled polymer composites, i.e. systems that consist of monomer (oligomer) and filler. The second approach enables one to perform a more detailed study of the processes that involve separate atom clusters and, hence, to tackle the mechanisms of interaction between polymer materials and solid fillers (dispersed, fibrillar or others). Thus, the characteristics of macromolecule chains may be explained also.

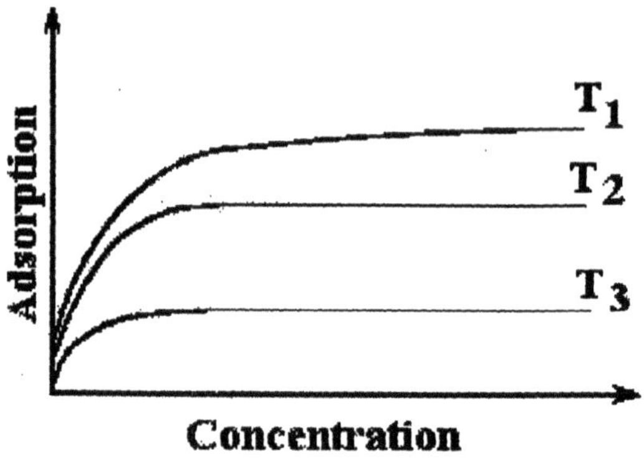

Figure 1. Relation between the quantity of adsorbed polymer A and its equilibrium concentration C at different temperatures

Adsorption studies and those of adsorption interactions between fillers and polymers prepared in advance, i.e. studies that employ the first approach, are well known. They tackle current processes that are essentially different from those phenomena, characteristic for low-molecule substances. As a result of the study of polymer adsorption on various fillers, a different character of the adsorption isotherms is established. The isotherms, however, express graphically the adsorption dependence on the polymer equilibrium concentration, for a constant temperature value [1, 9, 10, 15, 16] (Fig. 1).

Unlike the adsorption of low-molecule substances, a change of the location of the isotherms is possible for polymers when changing the temperature: in some cases the maximal adsorption is lower for higher temperatures (i.e. for $T_3 > T_2 > T_1$, Fig. 1), while in other cases it is higher, e.g. for $T_3 < T_2 < T_1$ in Fig. 1. Besides, changes during the isotherm shift can occur for a constant temperature - such are induction periods, curves, sharp increases and steep decreases of the quantity of adsorbed polymer, etc. These phenomena are explained by the theory of multilayer adsorption. Following this theory [1, 10, 16], the occurrence of curves in the isotherms of the adsorption of high-molecular substances may be regarded as a result of the change of the adsorption character. Such a change is expressed, most of all, by a shift to the adsorbent surface of macromolecule aggregates or supermolecular formations, being larger than those inherent to the previous sections of the curve. Using the initial sections of the adsorption isotherms, one can calculate the constants of adsorption equilibrium whose change is determined by the formation of adsorption links [1, 15, 17-19].

Most of the experimental and theoretical studies of polymer adsorption treat diluted polymer solutions [1, 20], since transition to concentrated systems yields additional difficulties. The change of the solution concentration causes a change of the conditions of inter-solution and surface interactions, as well as a change of the macromolecule conformation in the solution [4, 9, 12-14, 21-24]. The formation of macromolecule aggregates from the polymer is initiated right in the diluted solutions and hence, particles that differ from one another by shape and dimensions, adsorb under each concentration. Subsequently, there are changes in the contact conditions between molecules and molecule aggregates, as well as between molecules and the adsorbent surface; these changes induce a change of the adsorption layer structure [1, 25, 26]. For the adsorption of polymers from more concentrated solutions, for instance, it is established in [12, 13] that a shift to the filler surface takes place; this is not a shift of isolated polymer chains, but of molecule aggregates. This yields significant change in the molecule movability within the adsorption boundary layers and, subsequently, in the change of the whole complex of physico-chemical and mechanical properties. On the other hand, aggregation of macromolecules within the adsorption layer should also be considered. Such an aggregation is hampered by the effect of the surface where macromolecules are conformationally restricted [1, 3, 12, 16, 17, 21]. Besides, adsorption of polymers, even from diluted solutions, should never be treated as a process of the formation of monomolecule adsorption layer. These effects significantly hamper the disclosure of the basic laws of structure formation. Yet, it is well known that a significant deformation of the polymer chains occurs during the adsorption of polymers on the surface of a disperse filler. This means that the chain conformation changes [9, 25]. The effect is also related to a significant restriction of the molecule movability, resulting from the formation of adsorption links within the system [12, 13, 17].

The occurrence of such interactions is a result of the coordination between molecules, being accompanied by a redistribution of the electron density within the initial molecules [27-29] and by a surface restructurizing - see, for instance, [1, 3-13] where the occurrence of adsorbate “chemisorptive dissociation” is discussed. Finally, the donor-acceptor mechanism of adsorption of polymers should depend on the acid-base properties of all reagents in the system - polymers, solvents and adsorbent surfaces that are in charge of further interactions. In [30], for instance, a model system of acidic polymer (chlorinated polyvinyl chloride) and a basic filler (calcium carbonate) is studied. The authors establish that strong adsorption takes place for an acidic adsorption-admitting solvent (dichloromethane) and is entirely absent for a basic adsorption-hampering solvent (tetrahydrofuran).

In [31] the authors establish that a maximum adsorption of either stearic acid or polymethylmethacrylate (basic polymer) occurs from a nonpolar solvent (carbon tetrachloride), and that the adsorption decreases when acidity or basicity of the solvent increases. Depending on the magnitude of the acidic or basic character of the solvents [slightly basic (benzene) and acidic (methylene dichloride) and strongly acidic (chloroform) and basic (tetrahydrofuran)], a partial coverage to a maximum of two geometrical layers of either stearic acid or PMMA adsorption onto wollastonite is observed. This phenomenon may be linked with a bifunctional character (acid-base) of the wollastonite surface and with the competitive interactions in the system filler - main adsorbate (acid or base) - solvent (base or acid, respectively).

3. ACID-BASE INTERACTIONS IN SYSTEMS POLYMER - FILLER

If a polyamide-4 and polyamide-6 is filled with halogens of alkaline metals, there is a coordination linkage of metal ions with oxygen, and of halogen ions with nitrogen of the polyamide amide group [29]. A basic-acidic donor-acceptor interaction has been established for the adsorption of poly(oxyethylene) onto antimony sulphide, where the units of the poly(oxyethylene) are oriented towards the filler surface through their oxygen atom [25]. The same effect has been observed during the study of the adsorption of an acidic polymer [poly(vinyl acetate) and copolymers with acrylic acid]. For these filling substances, the quantity of adsorbed polymer grows significantly during a transition from an acidic to a basic filler. The transition order is established to be the following: SiO_2, Al_2O_3, TiO_2, Fe_2O_3, MgO, CaO [32]. [33] establishes the interaction between an acidic polymer (copolymer of the acrylic acid and ethylene) with a basic surface of an aluminium foil, based on the temperature of a surface adsorption-desorption of test molecules. The adhesion increases with an increase in the number of polymer acidic sites or of the aluminium foil basic sites. Foil wetting depends significantly on the chemical linkage between the two components of the system. The authors of [34] prove a strongly irreversible linkage between the basic poly(methylmethacrylate) on the surface of the acidic α-aluminium oxide by means of a hydrolysis of the lateral polymer ester groups through ion carboxylate groups. This proof, however, does not characterize any surface properties of the filler.

3.1 Acid-base interactions in polymerization and polycondensation filling

The clarification of the mechanism of adsorption interaction is a most important point of the processes of polymerization and polycondensation filling. Here the fore contact between the reagents should, without any doubt, significantly affect the structure of the final products, most of all when there are important changes in the reaction medium. They could be provoked by the acid-base interaction of the monomer (or oligomer) either with the solvent (if any) or with the filler. [7] emphasizes that a polymerization of the already coordinated monomers is necessary to increase the number of interacting particles in the system. [35] establishes that the polymerization rate of some monomers (methylmethacrylate, acrylamide) increases with the increase of the abilities of the solvent to form intermolecular links with the monomer, i.e. the rate increases with the increase of the solvent acidity. Due to the presence in the monomer of a proton-acceptor group with strongly expressed basic properties (the amino group here), there is an interaction between the group and the acidic solvent. This yields a redistribution of the electron density in the monomer molecule and, respectively, an increase of its reaction capabilities and branch formation, regarding once again the nitrogen atom. The decrease of the solvent acidity [36] decreases its donor-acceptor interaction with the basic monomer (acrylamide) where the branches along the amide group decrease, too. However, many solvents are able of such an intensive acid-base interaction with the monomer, so that the latter cannot be adsorbed and effectively linked with the filler [37]. Hence, it is necessary to choose a solvent to ensure interaction with the monomer, thus activating it, and providing for a real possibility of monomer adsorption on the filler surface. A reasonable compromise is needed. As established in [38, 39], the most "inclusion reaction" portion of the particles, should be weakly linked with the surface. It is proved that the existence of a strong adsorption link between an acidic monomer (acrylic acid) and a basic filler (aminated aerosil) or a strong adsorption link between a basic monomer (4-vinylpyridine) and an acidic filler (carboxylated aerosil) decreases the polymerization rate untill its termination. This is opposite to cases where a weak adsorption link between a monomer and a filler is observed - see, for instance [36, 40, 41] where the monomer is characterized by a strong surface mobility and by a high polymerization rate. [42] establishes that when the strength of the adsorption link between a set of vinyl monomers and the filler grows, the copolymers synthesized on the surface have few strong adsorption links. Moreover, the activation energy of polymer chain growth rises, too.

Many field studies have been performed recently. They aim at the clarification of the interaction between adsorbed monomers and disperse substrata; the mainly concern monomers, adsorbed from a gas phase, involving natural clay adsorbents [43-47] and metal oxides [6, 48]. [6, 45] establish that during monomer oligomerization, adsorbent activity correlates to its acidity. Note that the adsorbent acidity determines the mobility of the adsorbed molecules [46] and the direction of further interactions - oxidation, cross-linkage interlacing, cycling, surface - bulk crystallization [49], spontaneous or provoked polymerization, etc. The increased activity of the adsorbed monomers is explained mainly by the complexes, formed between them and other reagents, through a delocalization of unpaired electrons, and through further electron coupling along the chain [6, 27, 48, 50-53], i.e. through a so-called "non-dissociative adsorption". In most cases the activity of minerals, oxides and salts, used as fillers, is related to the existence of coordination-unsaturated atoms located on their surface that represent Lewis acidic sites, [6, 48, 54-60]. This activity is also due to the fact that those materials affect monomer molecule ordering within the adsorption layer [44, 47; 61-63]. [59], for instance, proves that a totally linear polybutadiene with a 1,4-configuration is formed during the polymerization of a 1,5-hexadiene in the presence of Lewis acids. [61] establishes that for the donor-acceptor adsorption interaction between the monomer basic groups (phenyl groups of diphenylsilanediol) and Lewis acidic sites, the interaction takes place on the surface of the studied filler (titanium cations of TiO_2). Note that the filler possesses a significant electron-acceptor ability and exercizes a polarizing effect over the monomer. On the other hand, there is an interaction between the monomer acidic functional groups (silanole groups) and the basic OH groups on the filler surface. The same authors prove that when decreasing the acidity of the used fillers (in the order TiO_2, Fe_2O_3, CoO_3 and CuO), filler interaction with the basic monomer groups decreases, too, in accordance with the decrease of the electron-acceptor capability and polarizing effect of the corresponding metal atoms. [60] proves the increase of the electrophility of a C atom from the imide group, due to its coordination with a Lewis acidic site. Hence, the electron pair increases the catalytic activity of the employed Mg-halogen complexes during an anion polymerization of ε-caprolactam. The separation of charged $C_4H_9(C_4H_8)_n^+$ particles as a result of the interaction between the monomer (isobutylene) and Zn^+ is proved in [48]. Hence, the induction of further intercluster polymerization is proved.

Similar results are obtained during the study of the adsorption interaction between thermosetting polymers and inorganic fillers. In these cases the specific interaction is accompanied by the participation of the functional groups of the curring agent in a non-dissociative adsorption that takes place on the filler surface, as is shown in [15, 17, 56]. [64] studies the synthesis of amino resins in the presence of complex-forming aprotic acids; the role of the latter consists in the activation of the carboxyl group at the expense of the interaction with the metal ion. During the study of the interaction between phenol formaldehyde oligomers and γ-Al_2O_3, it is shown that there is an activation of the oligomer molecules, adsorbed on the most active surface sites. Furthermore, the establishment of intermolecular links with other, still not adsorbed molecules, and the involvement of the latter in the adsorption process is also shown [15]. The same paper establishes that the better adsorption abillities of the resol phenol formaldehyde oligomers, as compared to the adsorption characteristics of the novolak oligomers, is determined by the presence of two types of functional groups, being active during adsorption.

From these studies we can conclude that a characteristic feature of the reactions that take place in the presence of solid fillers is their adsorption mechanism. The laws that these processes follow are determined by the character of the adsorption interactions between monomers (acidic or basic) and the surface active sites (correspondingly basic or acidic). These adsorption interactions determine also the degree of ordering, the structure of the adsorption layer and the direction of further reactions. At the current stage of the study we cannot present a unified mechanism that would describe the effects of adsorption and adsorption interactions between monomers and fillers, the substances being previously characterized from an acid-base point of view. Moreover, we are still not able to set forth how the adsorption mechanisms affect the development of further processes and the formation of the structure and properties of the obtained products. It is obvious that during a transition from a "gas-solid" to a "liquid-solid" system, adsorption interactions will become even more complex. Considering a "liquid-solid" system, one should start the interaction analysis by modelling the processes taking place on the filler surface. In other words, a model analog of the interactions between definite functional groups under the active participation of the interface should be sought, but a detailed description of the processes that take place on the filler surface and a definite clarification of their mechanisms still need further study.

The references considered in brief in this section treat adsorption interactions in filled polymer systems. They are linked with one another by accounting, most of all, for the change of the electron density of the participating compounds, for the delocalization and the bonds' conjugation, and as a result, for a certain orientation on the filler surface. All these effects yield an inevitable change in the thermodynamic parameters in a given "polymer/filler" system. Hence, the establishment of a change of the composite thermodynamic parameters should without any doubt be one of the proofs of adsorption interaction occurrence. Especially indicative in this respect is the adsorption heat which can be experimentally determined either by using indirect methods (getting adsorption isosters - isotherms at different temperatures, processing kinetic data etc.), or by employing a direct calorimetric method [1, 18, 19, 39, 65-69]. [65, 68], for instance, tackle the adsorption heat of poly(vinylchloride) and some surfactants from aqueous solutions. They establish that the heat increases with the increase of the filling degree. This phenomenon is explained by the prevailing character of the processes of polymer matrix disordering as a result of interactions with the filler. Despite the importance of a thermodynamic examination of filled polymer systems, there is still a gap in this field in the literature.

3.2 Spectroscopy of acid-base interactions

As noted, the formation of complexes through a charge transfer, as well as the formation of various donor-acceptor coordinate systems, are related to a change of the electron density of the initial donor and acceptor molecules, and to a subsequent change of the molecule equilibrium state [63, 70]. Information on such changes can be obtained by using spectroscopic methods of analysis, see, for instance, [71].

What is important when studying the spectra of various compounds is that the oscillations of one and the same chemical groups, being part of different molecules, are characterised by an absorption within one and the same spectrum area. A change of the oscillation position and intensity, and the disappearance or appearance of new absorption spectrum bands provide for various characteristics of the system structure: character of the functional group surrounding the medium, state of the molecules, formation of intramolecular and/or intermolecular linkages, etc. [5, 34, 53, 55, 72-74]. Employing this method, one can follow at a molecular and functional level the changes that occur in the polymer after the introduction of a filler. It is natural that these studies should be performed like the adsorption studies, i.e. first of all an interaction of previously prepared polymers and fillers should be analyzed, and secondly - interaction of monomers with fillers and further joint polymerization should be investigated. Using such a spectral method of study, one can obtain a complex and detailed idea of the physico-chemical interactions that take place in composite materials during their formation.

Polymer-filler interaction is identified in a number of papers through the change of the system infrared spectra. In 1963 [75] it was shown that considering the infrared spectrum of acetaldehyde adsorbed onto alumina, a significant shift of the carbonyl group band to the long wave range is observed as compared to the liquid monomer. The authors explain this effect by an ordering of the acetaldehyde molecules in the adsorption layer. This itself yields a monomer activation and a fast growth of the polymer chains, accompanied by the formation of a large quantity of amorphous rubber-like polyacetal. [76] employs FTIR spectroscopy and studies the interaction of allylbenzene with various oxide catalysts - acidic, basic and reduction-oxide. It seems that the authors have taken into account simulation of the surface "amphoteric" character of the interaction. Hence, they have included oxides of transition metals. In these cases it is firmly established that an interaction of the organic molecule with the inorganic substrata is governed by different mechanisms, depending on the surface acid-base character. However, not all the compounds are examined with respect to their type, strength and concentration. Lacking of such an analysis one is unable to link the character of the surfaces with their acidic or basic interaction by a definite allylbenzene parts (functional groups), even although the existence of a different coordinate intermediate during polymer activation and growth on the filler surface has been established.

The strong interaction between lateral ester groups of poly(methylmethacrylate) and the surface of α-alumina yields a hydrolysis accompanied by the formation of polymer ion carboxylate groups. Such a strong interaction is proved by using FTIR spectroscopy [34].

[73] establishes a complex formation between poly(methylstyrene) and $SnCl_4$ on the basis of the shift of the oscillation frequencies, and by the appearance of the new characteristic bands in the IR spectra. The same effect has been shown to hold for poly(vinylpyridine) and for different acceptors (halogens and metal chlorides). The authors of [77] study the change of the infrared spectra of acrylamide and poly(acrylamide), and of methylolacrylamide and poly(methylolacrylamide) in the presence of bentonite, MgO, SiO_2 and Al_2O_3. The formation of complex links between the monomer, polymer and the filler is proved on the basis of the appearance of new absorption IR bands. The participation of *N*-vinyl monomers in the formation of complexes is proved in [78], where the charge transfer is demonstrated to result from a donor-acceptor interaction, using spectral analysis. When studying the infrared spectra of polyurethane [79, 80] and of various types of rubber [81, 82] and accounting for the introduction of inorganic salts [83, 84], it is shown that an oscillation shift, as well as change of the intensity of the characteristic bands, take place. These effects are explained by the orientation of the macromolecules on the surface of the inorganic salt, resulting from a surface-polymer interaction. The authors of [72] study polyamide-6 filled with dehydrated $FeCl_3$. They establish a strong complex formation between the filler and the polymer NH-groups, on the basis of the change of the NH-group intensity.

Employing an infrared spectroscopy, a specific interaction is proved, i.e. between plane groups of poly(*m*-phenyleneisophtalamide) and salts that are present in the solution as ion pairs [83]. Following a donor-acceptor mechanism, paper [3] establishes the formation of chemisorption links between the amide group of poly(hexamethylene)sebacinamide and the free *d*-orbital of the metal atom of the filler (copper silicate). The results are obtained on the basis of a significant shift of the band of NH-group valence oscillations to the high-frequency range. [85] employs infrared spectroscopy and establishes that the curing of urea-formaldehyde oligomers is accompanied by intramolecular cyclization and hydrolysis. All these spectral effects are explained by the interactions occurring between the polymer and the inorganic substance. This yields a complex formation and ordering of the macromolecules near the surface of the filler particles.

Our investigation of acid-base interactions were performed by infrared study of the ε-caprolactam molecules adsorbed on the inorganic surface. As shown in Fig. 2, significant changes in the infrared spectrum of adsorbed ε-caprolactam can be seen and a surface effect is found even in the high-frequency range of the NH-vibrations (over 3000 cm^{-1}). There the adsorption in the presence both of γ-alumina and calcium phosphate leads to the disappearance of the absorption bands which are specific for hydrogen bonded amide groups at 3270 cm^{-1}, and to marked decrease of the first overtone of the deformation band at 3180 cm^{-1}. Note that, as compared with monomer in a condensed state, considerable low-frequency shift of the vibrational bands of carbonyl groups (in the range 1550-1800 cm^{-1}) can be seen in the adsorption. The CO-vibration shift is 45 cm^{-1} in the case of ε-caprolactam adsorbed on γ-alumina, and 30 cm^{-1} for adsorption on calcium phosphate. These effects can be attributed presumably to the interaction between functional groups of the monomer and the acidic sites on the inorganic surface. The stronger the adsorption interaction between monomer and filler, the bigger the shift of the absorption vibration band of the carbonyl groups.

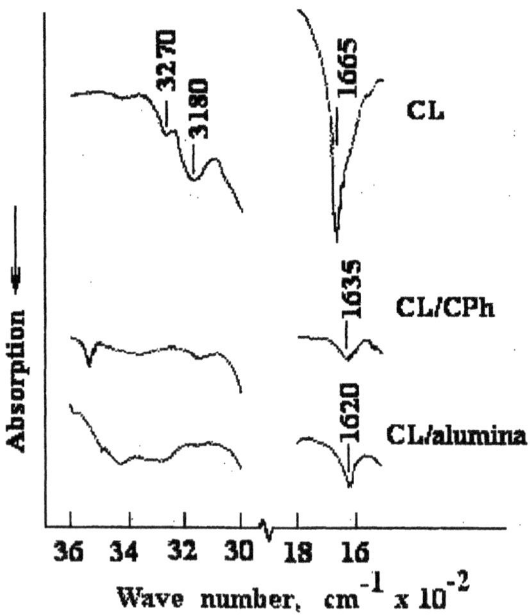

Figure 2. Infrared spectra of condensed ε-caprolactam (CL) and ε-caprolactam adsorbed on γ-alumina and calcium phosphate (CPh)

Spectral studies have been initiated in the last decade which consider not only the processes of on-filler adsorption of monomers, but also adsorption effects on the further polymerization. The authors of [39] note that the nature of the strong link occurring between monomers and fillers, is not entirely clarified. They outline the possibility of formation of two-centered complexes during adsorption. The authors assume that surface OH-groups, the carbonyl group and the nitrogen atom of the monomer molecule participate in that complex formation, resulting in the known orientation of the monomer molecule towards the solid body surface. The authors of [86] use infrared spectroscopy, and study ethyleneoxide adsorption and desorption from a gas phase onto inorganic oxides MgO and CaO, aiming to clarify the mechanisms of ethyleneoxide polymerization. They prove the adsorption of ethyleneoxide and the formation of a weak coordination link between unsaturated metal cations and the free electron pair of the monomer oxygen atom. However, in their study, the authors do not characterize the reagents, regarding strength and concentration of the existing sites, although they prove the existence of a coordination intermediate and its functioning as a point of polymer chain growth.

From these results we may conclude that the presence of unshared electrons and electron pairs in monomer and oligomer molecules (regarding heteroatoms - mostly oxygen and nitrogen), predisposes to a donor-acceptor interaction with the surface of the introduced filler. In other words, non-dissociative adsorption takes place, adsorption interactions occur in the systems as a result, and this leads to the orientation of the adsorbed molecules. Such changes of the system can be registered by employing a spectral analysis, for instance by infrared spectroscopy. During the course of further development of the process on the surface, a mutual orientation between monomers and fillers can cause various processes, leading to the surface changes - oxidation, cross-linking, chain branching, etc. The mechanisms of the phenomena that take place at the interface, i.e. within the boundary layer, are still not fully explained. Different authors assume a possible orientation of the components on the solid body surface (i.e. on the filler surface) and a certain local redistribution of the compound concentration. Sometimes it is assumed that the presence of uncompensated valences, being a characteristic for a number of oxides and salts, may cause some polar and even ionic interactions, and may yield to the formation of various coordination links, as well as transitions through different intermediate compounds. Regarding all these interactions, we must note an extremely important point - the presence of a solvent film (in particular, water), adsorbed on the surface. Some authors assume that such a film could screen the active reaction sites [5, 22, 87] while others assert quite the contrary - that such a film could stimulate the formation of new links (hydrogen, ionic, donor-acceptor ones, etc.) [53, 64, 78, 84, 88].

Some researchers suppose that when introducing inorganic salts, the polymer could be presented as a copolymer with blocks of polymer chains and salt molecules, regarding ionomeric structures which building units have different mobility [60, 88-90]. No surface acidic and basic properties of the reagents are considered here. The authors of [83] assume that the salts introduced in the solution have an ionic pair form; the realized interaction between the ionic pair dipole and the polymer functional groups helps the polarization of the macromolecule, and provides for an interaction between groupings of polymer chains and solvent molecules.

4. STUDY OF A MODEL ADSORPTION SYSTEM

An important first stage in the study of the interaction between compounds, participating in the reacting system, is the examination of the monomer adsorption on the filler surfaces. The resulting adsorption layers can significantly change the interaction conditions, and subsequently the properties of the whole system. Hence, the study of the laws of formation, structurizing and properties of adsorption layers turns out to be a basis for the analysis of the properties of these systems; adsorption kinetics provides important information on the processes which occur.

Regarding the literature survey of the results on interactions that take place in a heterogeneous system, we consider that the interaction between the inclusion compounds is of specific importance for the further development of the processes. As shown in [53, 91-93], heterogeneous systems can be divided into two basic types: i) systems with weakly developed interactions between phases and ii) materials where the phases are chemically grafted. Note that [93] confirms the fact that when the intermolecular distance attains a value, smaller than 4 Å, chemical adsorption takes place inevitably, owing to the molecular contact between the polymer and the surface. [93] does not study the acid-base state of the system components at all.

In the stage of preparing filled polymers by employing a polycondensation or polymerization filling, the interaction between monomers (oligomers) and the filler surface in use turns out to be a decisive factor; this interaction has an adsorption character. As proved in Chapter 1, the occurrence of adsorption linkages and the elucidation of the mechanism of adsorption interaction depend to a certain extent on the acid-base properties of the all participating reagents. This means that the interaction depends on the character of the monomer (oligomer) functional groups and as well on the presence of acidic sties with various strengths and concentrations on the filler surface.

Due to the extreme complexity of the occurring processes, we propose a detailed study of the interactions that take place in composite materials, based on the employment and analysis of model systems. Following the definitions of Drago [94, 95] and Tanabe [96], solid acids are the often used fillers for polymers. Hence, some oxides and salts often used as fillers for polymer materials, are used as solid acids in the model studies. Such are, for instance, SiO_2 - quartz, TiO_2 - rutile, γ-Al_2O_3, $BaSO_4$ and $Ca_3(PO_4)_2$. To study the adsorption acid-base interactions on the surface of these solid acids, it is necessary to choose a monomer with basic properties that possesses functional groups to model common polymers. These are carbonyl and amide containing monomers. Despite that, a study of the interactions between filler and thermosetting oligomers surface is too hard to accomplish, since these compounds are concentrated solutions. Even obtaining diluted solutions of these reagents is practically impossible. Based on these theoretical assumptions and practical difficulties, we study acrylamide monomer as a model compound.

The methods of obtaining adsorption isotherms are described at the beginning of the present chapter. Figures 3 and 4 show the experimentally determined isotherms of the adsorption of acrylamide on various inorganic compounds. The figures show that the adsorption of acrylamide on different surfaces decreases gradually untill saturation of all surfaces is attained.

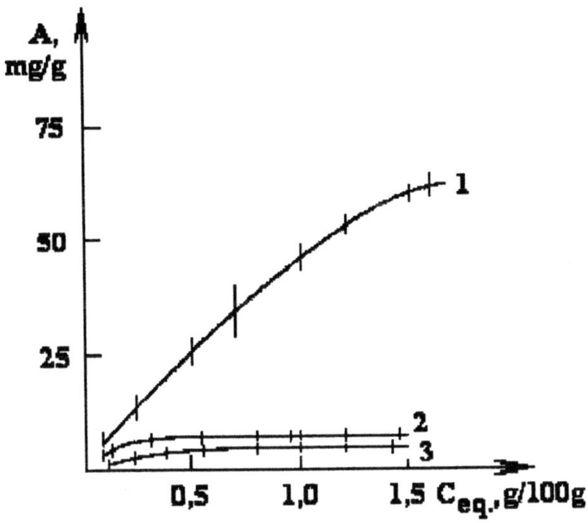

Figure 3. Adsorption isotherms of acrylamide on TiO_2 (*1*), SiO_2 (*2*) and $BaSO_4$ (*3*) surfaces

Suppose that the specific surface of the adsorbent is determined by using a certain physical method, i.e. a low temperature adsorption of nitrogen, for instance. Suppose also that the values of A_m are calculated, using the corresponding adsorption data. We can calculate the area w occupied by the adsorbate molecule in a densely packed monolayer:

$$w = \frac{s}{A_m \cdot N_A}, \qquad (2.13)$$

where N_A is the Avogadro's number, s is the adsorbent specific surface. Using these results, we can graphically present the effect of the surface character on the packing density of the adsorbed molecules.

On the other hand, using the geometrical sizes of the acrylamide molecule and supposing that a totally packed monomolecular adsorption layer is formed, we can calculate the theoretical area on each surface, occupied by the molecules; the molecules are supposed to be perpendicular to the adsorbent surface. We have calculated the dimensions of the acrylamide molecule, regarding the covalence radii of all participating atoms and the subsequent angles of inter-atom links. We obtained the molecule possible maximal length - $5.47.10^{-10}$ m. The theoretically admissible maximal area thus calculated that is occupied by a single acrylamide molecule, is w_t = $19.2.10^{-20}$ m^2, while the maximal theoretical adsorption $\Gamma_{m,t}$ is 0.61 mg/m^2. Hence, using the inorganic compounds specific surface determined by data on nitrogen low temperature adsorption, through formula (2.13) we can theoretically calculate the maximal possible quantity of acrylamide, adsorbed in the monomolecular layer covering the subsequent surfaces $A_{m,t}$. The results obtained by the experimental study of acrylamide adsorption and by theoretical calculating the area w_t and the maximal adsorption in a monolayer $A_{m,t}$, are given in Table 1.

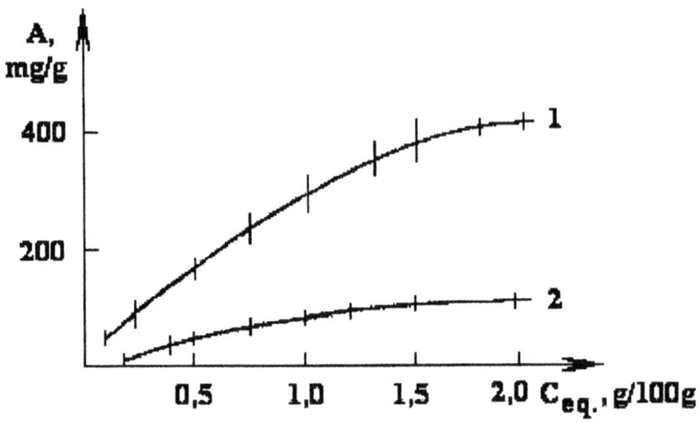

Figure 4. Adsorption isotherms of acrylamide on γ-Al_2O_3 (*1*) and $Ca_3(PO_4)_2$ (*2*) surfaces

It is seen from the table that the experimental values of the maximal adsorptions of acrylamide on inorganic compounds are different for different surfaces under study and can be ordered in rows as follows:

- per gram adsorbent

$$BaSO_4 < SiO_2 < TiO_2 < Ca_3(PO_4)_2 < \gamma\text{-}Al_2O_3$$

- per adsorbent unit surface

$$SiO_2 < Ca_3(PO_4)_2 < BaSO_4 < \gamma\text{-}Al_2O_3 < TiO_2$$

The results from the briefly presented studies that concern the adsorption of acrylamide on various inorganic surfaces show the presence of a plateau on the isotherms, i.e. the occurrence of a surface saturation, while the calculations prove that there is polymolecular (multilayer) adsorption. Analyzing these data, however, we cannot give answers to other important questions - what is the interaction nature and what is the orientation of the adsorbate on the adsorbent surface. A single study of monomer adsorption on an inorganic surface, from an acid-base point of view, is not able to provide full and exact information on the character of the adsorption processes. To study the surface compounds formed during adsorption, one needs to employ other physical methods which could effectively characterize the "adsorbate-adsorbent" interactions; here infrared spectroscopy is most valuable enabling one to analyze the adsorption of distinct adsorbate molecules on separate adsorption sites.

Table 1. Experimental and theoretical results of the adsorption of acrylamide on various surfaces

Filler Parameter	SiO_2	TiO_2	γ-Al_2O_3	$BaSO_4$	$Ca_3(PO_4)_2$
Specific surface, S, m^2/g	9	15	155	3	66
Maximal theoretical adsorption, $A_{m,t}$, mg/g	5.54	9.22	95.32	1.84	50.43
Maximal experimental adsorption, $A_{m,e}$, mg/g	7.55	65	434	4.5	96.5
Maximal experimental adsorption, $\Gamma_{m,e}$, mg/m^2	0.84	4.33	2.8	1.5	1.46

Our conclusions of how surface interactions affect the properties of the final polymer composite materials are in accordance with the ideas of Fowkes and his research group. Note that surface effects hold during composite preparation, and during further material processing - filling, curing, restructurizing, etc. Tribute to Fowkes is paid by the generalized paper [97], dedicated in his honour. However, this is only an initial step into the research of a complex organic-inorganic system; there is still more to be done. However it is clear that the study of such systems could provide information that would more or less elucidate the processes that take place on the surface of the inorganic filler, and disclose how adsorption processes affect the properties of the final products - polymer composite materials.

4.1 Study of Surface Compounds of a Model System by Infrared Spectroscopy

As already noted, the strength of the links between the solved substance and the adsorbent surface divides systems into two formal groups - those characterized by *chemical* interactions, and those characterized by *physical* interactions. As is established during adsorption studies, different types of interactions can be observed - from the weakest ones employing van der Waals forces, to the strongest, real chemisorption. The shape of the adsorption isotherms provides only qualitative information on the interaction. However, an exact analysis of the type of linkages between the adsorbate and the surface can be made when studying the adsorption layer on a micro level, for instance by using infrared spectroscopy.

The specific character of adsorption interactions is usually determined by the chemical nature of the interacting substances. Besides, the chemisorption can be preceded by a physical adsorption. In both cases adsorbate-adsorbent interactions take place accompanied by changes of the compound electron density, and by a subsequent change of the oscillations of the surface groups or adsorbed molecules. This yields a shift of the characteristic absorption bands and even to the disappearance or appearance of new bands in the infrared spectrum.

Besides the double π-link, there are two functional groups in the acrylamide molecule, where the heteroatoms (N and O) have unshared electron pairs:

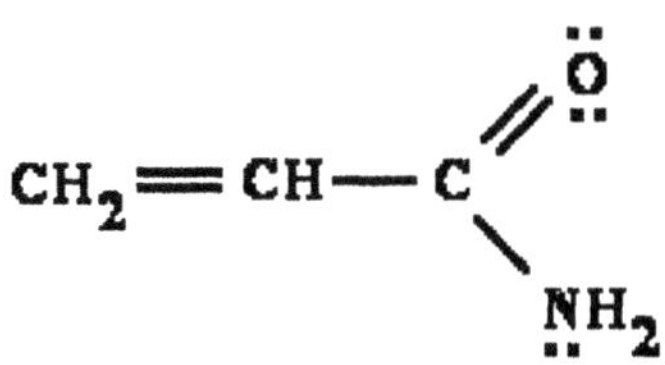

Using the unshared electron pairs, a donor-acceptor interaction can be accomplished besides the π-link, with electron donation and proton acceptance taking place. Note that the characteristic oscillation bands in the infrared spectrum significantly change their character in this case. In adsorption interaction, bands of valence oscillations of the carbonyl group [ν(C=O)] and of the deformation oscillation of the amide group [δ(NH$_2$)] are most sensitive; we concentrate the change in these bands.

Table 2 presents the absorption bands ν(C=O) and δ(NH$_2$) during the adsorption of acrylamide on different inorganic surfaces.

Table 2. Characteristic infrared bands of acrylamide absorption

Wave number, cm^{-1}	**Acrylamide**	**Acrylamide adsorbed on**				
		SiO_2	**TiO_2**	**γ-Al_2O_3**	**$BaSO_4$**	**$Ca_3(PO_4)_2$**
δ(NH$_2$)	1610	1610	1610	1610	1623	1622
ν(C=O)	1677	1677	1677	1677	1650	1677

The table shows that the spectra of the acrylamide that is adsorbed on SiO_2, TiO_2 and γ-Al_2O_3 do not differ from those of the initial acrylamide. During the adsorption of acrylamide on $BaSO_4$ and $Ca_3(PO_4)_2$, however, significant change is observed, i.e. splitting of the absorption bands of the NH_2 groups into two components. Besides, a significant shift of 27 cm^{-1} of the carbonyl band to the low frequency range is observed in the spectrum of $BaSO_4$.

The split of the bands of the NH_2 groups can be caused by the interaction of the $BaSO_4$ and $Ca_3(PO_4)_2$ aprotic acidic sites with the acrylamide; both types of amide groups - bonded and non-bonded [72, 98, 99], are present here. Such an interaction is accomplished by a transfer of the nitrogen unshared electron pair to the electron acceptors located on the inorganic surface where coordination bonded compounds are formed. A similar effect [split of the oscillation band $\delta(NH_2)$] is observed and reported in papers [77] and [100] where the authors suppose that this effect is determined by the acid-base interaction of the acrylamide with the studied aluminosilicates.

The observed shift of the oscillation bands of the carbonyl group during adsorption on $BaSO_4$ can be explained by the strong complex formation between the C=O group and the aprotic acidic sites - in that case they could be Ba^{2+} [69]. The whole carbonyl group participates in the interaction in this case

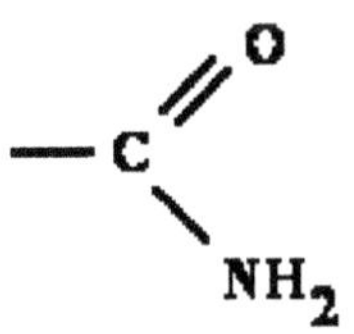

This follows from the changes occurring in both characteristic bands - $\nu(C{=}O)$ and $\delta(NH_2)$. Note that the acrylamide state on the $BaSO_4$ surface differs from that on the $Ca_3(PO_4)_2$ surface.

The analysis of the infrared spectra shows clearly significant differences in the state of the acrylamide adsorbed, on the one hand, on the surface of the studied oxides, and on the other hand, on the surface of the salts. The difference in the spectral behaviour of acrylamide adsorbed on $BaSO_4$ and $Ca_3(PO_4)_2$ is also established. In all the studied systems, the differences of the character of interaction between the inorganic compounds and the acrylamide are aroused due to the different nature of the acidic sites located on the inorganic material surface.

4.2 Infrared spectroscopy of test molecules

The specificity of the acrylamide adsorption interaction with various inorganic substances cannot be characterized by a single registering of the subsequent infrared transition spectra of the systems adsorbate/adsorbent. To perform a more detailed study of the character of adsorption interaction, one should employ other methods of infrared spectroscopy that deal with the analysis of the surface compounds. Note the extreme importance of the infrared analysis of the behaviour of various test compounds adsorbed (nitrogen containing compounds, such as pyridine [101], heavy water - D_2O, etc.). Depending on their chemical structure and properties, they interact with the acidic surface sites by following different mechanisms. Our interest is directed towards the establishment of spectral differences in the process of formation of hydrogen-bonded modes or during substitution for test molecules. The hydrogen-bonded or substituted modes arise from the display of surface proton donating or electron accepting properties. Such modes are formed during transfer of an electron charge to or from the surface, and this transfer is due to the surface acidic properties - protonic or aprotic.

A further study of the intermolecular interactions in acrylamide/$BaSO_4$ and acrylamide/$Ca_3(PO_4)_2$ systems is performed by the spectral analysis of the interaction of these systems with test compounds. We study the infrared spectra of $BaSO_4$ and $Ca_3(PO_4)_2$ with acrylamide adsorbed on their surfaces, as well as further interactions of these compounds with heavy water and pyridine. Substitution of hydrogen from surface hydroxyl groups by deuterium is performed by using D_2O vapor phase. The study aims at attaining a shift of the oscillation band of these groups from the range of weak transmission to the range of strong transmission. In addition, we aim to get a weaker scattering of the infrared radiation.

For both compounds - $BaSO_4$ and $Ca_3(PO_4)_2$, the exchange of deuterium takes place extremely easily - a wide band of absorption of the valence oscillations of groups O-D is formed at 2400 - 2700 cm^{-1}, after only 5 minutes of contact. For $BaSO_4$ this band remains for a long period of evacuation - even over 8 h (spectra 4 and 5 in Fig. 5). This confirms the stability of the deuterium exchange. Spectra 4 and 5 in Fig. 5 show that the acrylamide does not screen the hydroxyl proton sites on the surface of $BaSO_4$ - this, however, arises from the easy accomplishment of the deuterium exchange.

We obtain different results when considering an acrylamide/$Ca_3(PO_4)_2$ system and subsequent deuterium exchange. It is clearly seen from spectrum 5 in Fig. 6 that the evacuation that takes place after an interaction with D_2O yields a significant decrease of the intensity of the band of D-O group valence oscillations at 2400 - 2700 cm^{-1}. This shows that the deuterium exchange is not stable; this may be explained by the interaction between the acrylamide and different-in-nature acidic sites located on the surface of $Ca_3(PO_4)_2$, but the problem still needs further elucidation.

The assessment of the degree of screening of aprotic acidic sites, located on the surfaces of $BaSO_4$ and $Ca_3(PO_4)_2$, is made by studying the spectra of pyridine taken from a gas phase and adsorbed on the surfaces. Pyridine is chosen for this purpose, since it is a convenient and well known indicator for identifying the type of acidic sites located on solid acid surfaces. Moreover, pyridine is known to interact with acidic sites of different nature and its spectra are quite characteristic and illustrative.

Fig. 5, spectrum 6 shows that after the adsorption of pyridine processed with acrylamide and D_2O, a strong band of absorption at 1450 cm^{-1} appears in the $BaSO_4$ spectrum. The band may usually be related to the formation of a coordination linkage between pyridine and surface aprotic sites. This fact shows that not all aprotic sites, located on the $BaSO_4$ surface, participate in the interaction with the acrylamide. Pyridine forms an arbitrary weak coordination link with these sites, since it totally desorbs for a single hour of specimen evacuation - disappearance of the absorption band is observed in Fig. 5, spectrum 7, at 1450 cm^{-1}. Simultaneously, the absorption band at 1450 cm^{-1} is absent from spectrum 6 in Fig. 5, which is characteristic for protonized pyridine. This means a lack of active Brønsted acidic sites that could interact with the pyridine through hydrogen bonding.

To interpret the spectra of the $BaSO_4$ adsorption modes of acrylamide more fully, we use a method of comparison of these spectra with spectra of the systems "acrylamide/typical proton-donating medium" and "acrylamide/typical aprotic agent". The boric acid was used as a proton-generating medium, while aluminium chloride stands for an aprotic agent. Mechanical mixtures of acrylamide plus $AlCl_3$ and acrylamide plus H_3BO_3 are used to register the corresponding spectra. Such mixes are employed to avoid the solvent effect. Fig. 7 shows that the spectrum of acrylamide-protonic acid (H_3BO_3) (spectrum 4), does not differ from that of the initial acrylamide - spectrum 1. This means that the acrylamide does not display a capability of interacting with the free protons through the carbonyl group oxygen or through the amide group nitrogen. At the same time, spectrum 6 of acrylamide mixed with the strong aprotic acid $AlCl_3$ undergoes significant change in the range, characteristic for these groups, and entirely corresponds to the spectrum of the acrylamide adsorbed on $BaSO_4$ (spectrum 2). In both cases the band of the deformation oscillations of the amide group $\delta(NH_2)$ splits into two components - at 1610 and 1575 cm^{-1}, while the band of the carbonyl group shifts to lower frequencies. If we compare the spectrum of acrylamide, adsorbed on $BaSO_4$ (spectrum 2) and that of the $AlCl_3$/acrylamide mixture (spectrum 6), we can assess the quantitative correspondence between the changes that occur during the contact between the acrylamide and $AlCl_3$ and $BaSO_4$. Thus it is established that the aprotic sites located on the surface of $BaSO_4$ are responsible for the $BaSO_4$ - acrylamide interaction.

However, the relation of the pyridine with the system "acrylamide - $Ca_3(PO_4)_2$" is quite different. From fig. 6 spectrum 6 it is seen that a very weak band of absorption at 1540 cm^{-1} occurs. This is characteristic for the pyridine ion and proves the presence of Brønsted acidic sites on the surface. The system evacuation yields a total pyridine desorption (spectrum 7 in Fig. 6), i.e. pyridine ion linkage with the protonic acidic sites disappears. The existence of such weak protonic acidic sites could explain the split-up of the deformation oscillation band $\delta(NH_2)$ of the acrylamide adsorbed on $Ca_3(PO_4)_2$ - Table 2. Such a split-up occurs, because of the presence of acrylamide molecules that are bonded to the surface either by a protonic or by an aprotic linkage.

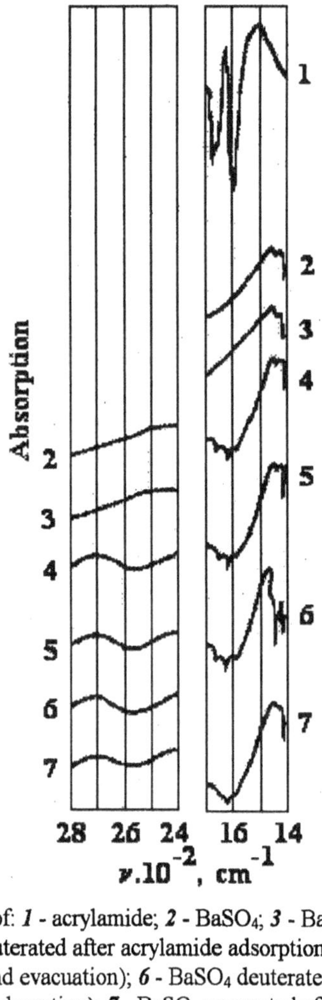

Figure 5. Infrared spectra of: ***1*** - acrylamide; ***2*** - $BaSO_4$; ***3*** - $BaSO_4$ thermally treated at 413 K for 3 hrs; ***4*** - $BaSO_4$ deuterated after acrylamide adsorption; ***5*** - $BaSO_4$ deuterated after (acrylamide adsorption and evacuation); ***6*** - $BaSO_4$ deuterated after (acrylamide adsorption and subsequent pyridine adsorption); ***7*** - $BaSO_4$ evacuated after (acrylamide adsorption and subsequent pyridine adsorption and evacuation)

The analysis of "adsorption system-test compound" spectra confirms the supposition that the interaction differences between inorganic compounds and the acrylamide are aroused due to the different nature of the existing acidic sites. Analyzing the system "acrylamide - $BaSO_4$", we found that the interaction involves the surface aprotic acidic sites, while the acrylamide adsorption interaction with $Ca_3(PO_4)_2$ is realized by a protonic and as well by an aprotic linkage.

These studies show that the acrylamide can interact during adsorption with the aprotic acidic sites located on the surface of the solid acids $BaSO_4$ and $Ca_3(PO_4)_2$, at the expense of the carbonyl group oxygen in the carbonyl group and the amide group nitrogen.

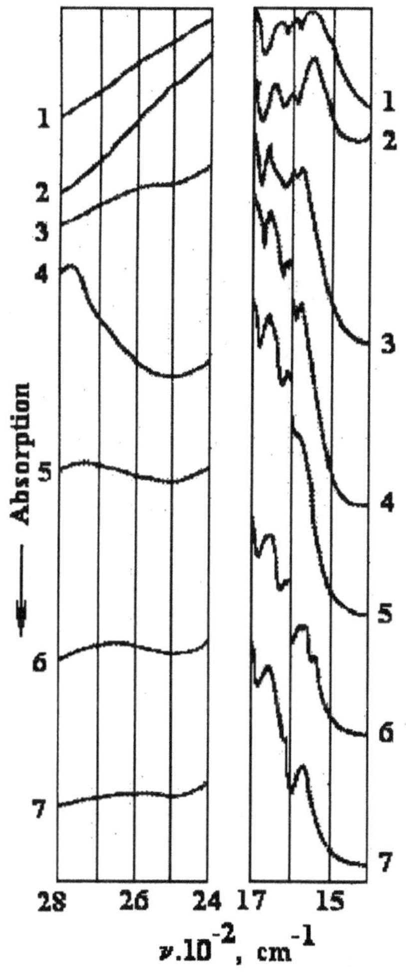

Figure 6. Infrared spectra of: ***1*** - acrylamide; ***2*** - $Ca_3(PO_4)_2$ with adsorbed acrylamide; ***3*** - $Ca_3(PO_4)_2$ with adsorbed acrylamide treated at 733 K for 3 hrs; 4 - $Ca_3(PO_4)_2$ deuterated after acrylamide adsorption; 5 - $Ca_3(PO_4)_2$ deuterated after (acrylamide adsorption and evacuation); 6 - $Ca_3(PO_4)_2$ deuterated after (acrylamide adsorption and subsequent pyridine adsorption); 7 - $Ca_3(PO_4)_2$ deuterated after (acrylamide adsorption, subsequent pyridine adsorption and evacuation).

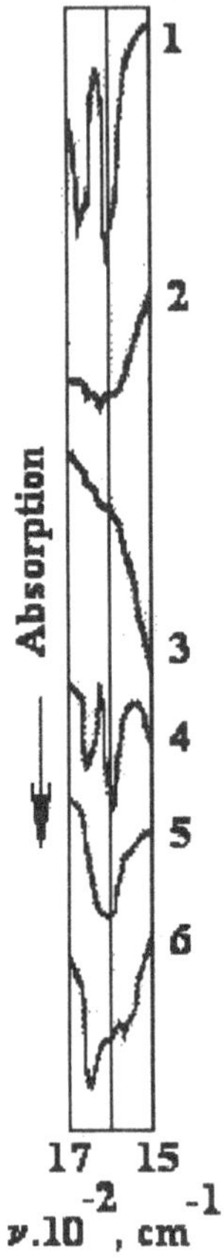

Figure 7. Infrared spectra of: ***1*** - acrylamide; ***2*** - acrylamide adsorbed on $BaSO_4$; ***3*** - boric acid; ***4*** - mixture of boric acid and acrylamide; ***5*** - $AlCl_3$; ***6*** - mixture af $AlCl_3$ and acrylamide

4.3 Fourier Transform Infrared Spectroscopy of Model Systems

Recently, the Fourier transform infrared spectroscopy has been successfully employed to study the surface compounds that are formed. Hence, our further study of the specificity of the intermolecular interactions that take place in acrylamide-inorganic surface systems continues with data from Fourier spectroscopy. We use a diffusion-scattered radiation device to record spectra; in this way we can obtain spectra of air-dry specimens without using an immersion liquid. We can also get information on the state of the adsorbed phase, using not only the adsorbed compound spectrum, but also the spectrum of the interacting surface sites active in adsorption. An additional advantage of using diffuse radiation is that the specimen is not heated by the diffuse-scattered flux of infrared rays. The figures that follow show Fourier spectrograms where the absorption *A* is given in relative dimensions - $ln(I_0/I)$, and I_0, I are the intensities of the incident and reflected light, respectively.

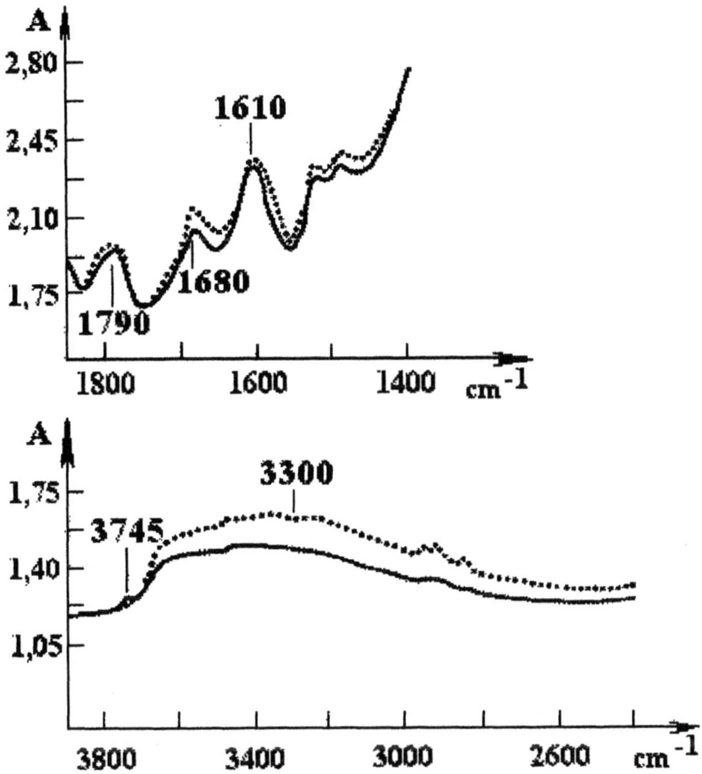

Figure 8. Fourier transform infrared spectra of SiO_2 before (solid line) and after (dotted line) acrylamide adsorption

Fig. 8 gives Fourier spectra of SiO_2 specimens that are obtained before and after adsorption of acrylamide; the acrylamide adsorption leads to the disappearance of the band characterizing the free silanol groups (ν_{OH} = 3745 cm^{-1}) on the SiO_2 surface. Band disappearance is accompanied by an increase in the absorption intensity of hydrogen-bonded associates near to the 3300 cm^{-1} range. The value of the band maximum shift of the free silanol groups on the surface is determined by the energy of the hydrogen linkage formation and is equal to $\Delta\nu_{OH}$ = 450 cm^{-1}. The spectrum of the adsorbed acrylamide, registered in the ranges of valence oscillation of the carbonyl group ν(C=O) and deformation oscillations of the amide group δ (NH_2), hardly differs from that of the acrylamide in the condensed phase (see Table 2); this accords with the spectral data obtained by using transmission technique, presented earlier.

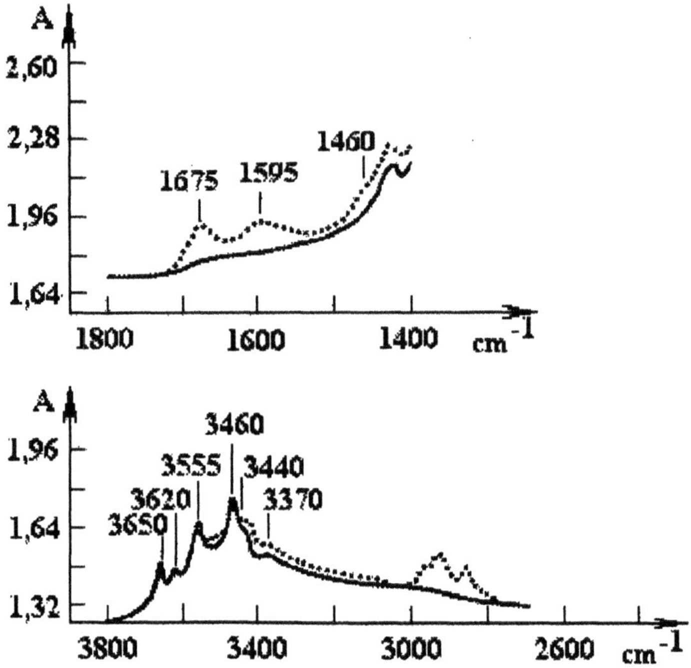

Figure 9. Fourier transform infrared spectra of γ-Al_2O_3 before (solid line) and after (dotted line) acrylamide adsorption

The adsorption of acrylamide on the surface of γ-Al_2O_3 (Fig. 9), unlike the adsorption on SiO_2, does not yield a change of the Fourier absorption spectrum of the valence oscillations of the surface hydroxyl groups. However, a slight decrease of the band intensity at 3555 and 3460 cm^{-1} is observed, which might be caused by a weak protonic interaction with the surface acidic sites. The bands of absorption of non-hydrogen bonded hydroxyl groups at 3650 and 3620 cm^{-1} remain unchanged. Bands of absorption at 1595 and 1675 cm^{-1} are observed in the spectrum of the adsorbed acrylamide. It must be noted that, unlike data from the transmission method (see Table 2), the band of deformation oscillation of the amide group δ(NH_2) is shifted from 1610 cm^{-1} to 1595 cm^{-1}. Such a shift in the low-frequency range can be explained by the coordination of the NH linkages with the inorganic surfaces that takes place at the expense of the nitrogen atom and by the weak amine hydrogen-surface linkage. The latter is due to the aprotic link localization at the nitrogen atom and to a corresponding weakening of the hydrogen atom. Thus, the intermolecular interaction of the acrylamide with γ-Al_2O_3 can be treated as arising from a donor-acceptor interaction between the amide groups and the surface protonic and aprotic sites, the latter being coordination-unsaturated aluminium atoms. The decrease of band intensities at 3555 and 3460 cm^{-1}, considering valence symmetric and asymmetric oscillations of the coordination-bonded hydroxyl groups, also confirms this conclusion.

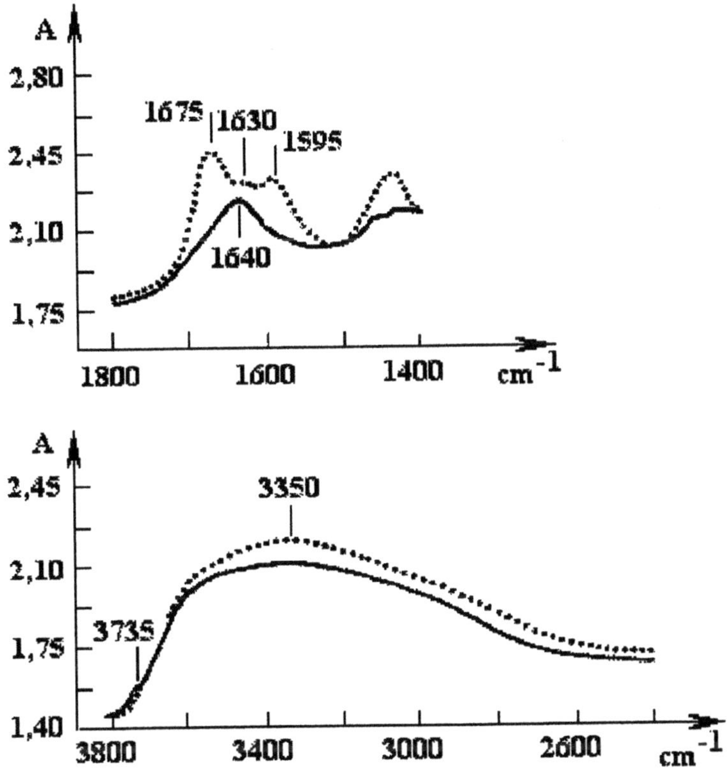

Figure 10. Fourier transform infrared spectra of TiO_2 before (solid line) and after (dotted line) acrylamide adsorption

Fig. 10 presents the Fourier spectra of TiO_2 before and after adsorption of acrylamide; the adsorption of acrylamide is accompanied by a disappearance of the weak absorption band at 3735 cm^{-1}. This band is related to the valence oscillation of the surface free hydroxyl groups, whereas the oscillation is accompanied by simultaneous increase of the absorption of hydrogen-bonded associates in the 3350 cm^{-1} range. These two effects prove the formation of a hydrogen linkage between the acrylamide and the surface free titanol groups. The deformation band of the amide group of the acrylamide, as in the case of γ-Al_2O_3, appears at 1595 cm^{-1}. More than that, a band appears at 1630 cm^{-1}, too, i.e. a split of the absorption band of the $\delta(NH_2)$ amide group takes place. This attests to the presence of donor-acceptor interaction with the surface protonic and aprotic sites (i.e. with coordination-unsaturated titanium atoms).

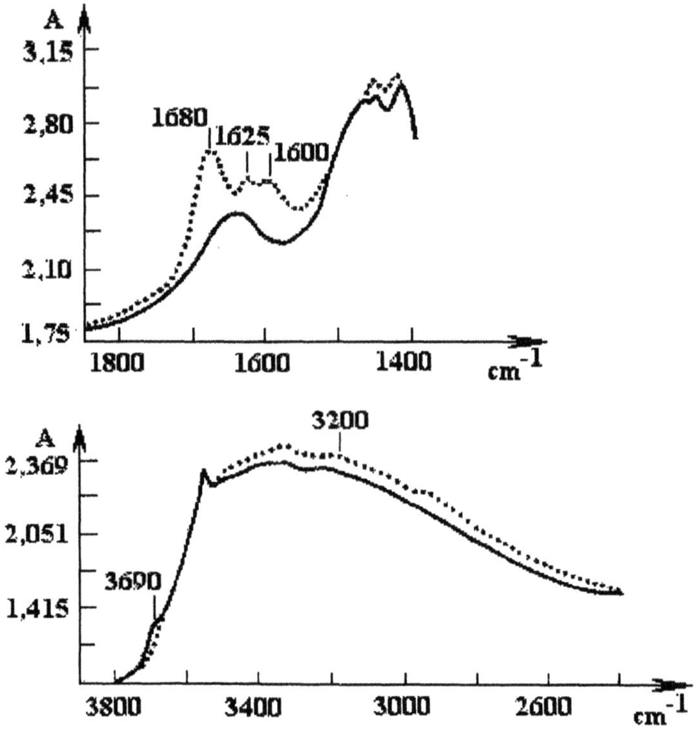

Figure 11. Fourier transform infrared spectra of $Ca_3(PO_4)_2$ before (solid line) and after (dotted line) acrylamide adsorption

Under adsorption of acrylamide on the surface of $Ca_3(PO_4)_2$ (Fig. 11), a qualitative similarity with Fourier-spectral acting of the adsorption on TiO_2 is established when comparing the band type at 1625 and 1600 cm^{-1}. The clearly expressed split of the band of the amide group into two components may also be due to the interaction with different-in-character acidic sites or to the fact that not all acrylamide molecules participate in the interaction with surface aprotic sites - coordination-unsaturated atoms Ca^{2+}. Together with the specifically oriented "molecular" form of the adsorption, we observe that the absorption band of the free hydroxyl groups on the surface disappears. These groups are in the form

$$\equiv P-OH$$

arising at 3690 cm^{-1}. Moreover, increase of the absorption intensity of hydrogen-bonded associates is also registered, and the intensity gains a maximum value in the 3200 cm^{-1} range. This effect shows that in this case the intermolecular interaction yields hydrogen linkage, too.

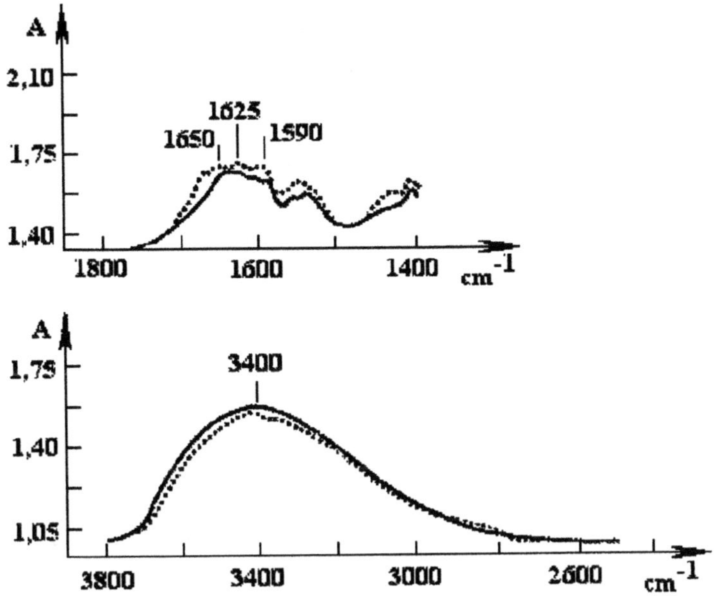

Figure 12. Fourier transform infrared spectra of $BaSO_4$ before (solid line) and after (dotted line) acrylamide adsorption

Fig. 12 presents Fourier-spectra of $BaSO_4$ before and after acrylamide adsorption. The change of the adsorbate spectrum is expressed by a split of the band $\delta(NH_2)$, as for TiO_2 and γ-Al_2O_3. Moreover, referring to the spectrum of the non-adsorbed acrylamide, we simultaneously observe a significant low-frequency shift to 1650 cm^{-1} of the absorption band of the carbonyl group $\nu(C=O)$ (see Table 2). These effects can be explained by the monomer coordination mechanism of interaction that is realized by means of the interaction of a carbonyl group through the aprotic sites located on the $BaSO_4$ surface (such are, for instance, the Ba^{2+} ions) as well as by means of the amide group nitrogen linkage.

Based on data from infrared Fourier spectroscopy, we may conclude that the surface adsorbent acidity significantly affects the character of adsorption interactions taking place on the interface. It is established that the intermolecular acrylamide - inorganic surface interactions are realized as follows:

- interactions with SiO_2 - following the mechanism of hydrogen linkage formation, since such a surface constitutes protonic acidity (Brønsted acidity) due to the presence of silanol groups on it;

- interaction realized as a coordination linkage between acrylamide molecules and aprotic sites located on the $BaSO_4$ and owing to the presence of coordination unsaturated metal atoms (Lewis acidity). The coordination linkage of acrylamide with the $BaSO_4$ surface develops in the presence of amide and carbonyl group. It is accomplished through an interaction with the aprotic acidic sites, thus confirming the results obtained from the transmission methods.
- a simultaneous interaction for γ-Al_2O_3, TiO_2 and $Ca_3(PO_4)_2$, employing the hydrogen linkage and a coordination linkage. The effect is due to the following characteristics of γ-Al_2O_3, TiO_2 and $Ca_3(PO_4)_2$ surfaces - presence of hydroxyl, titanol and $\equiv$P—OH groups (Brønsted acidity) and of coordination-unsaturated metal atoms - Ti^{4+}, Al^{3+} and Ca^{2+} (Lewis acidity).

The formulated and experimentally proved ideas for tackling with different types of interactions depending on the state of the studied surfaces confirm the gains to the knowledge in the field - see, for instance, [97, 101].

A mechanism of interaction between the acrylamide and the surfaces of inorganic substances with known acidity can be designed, based on the results of adsorption studies. It is established that such a mechanism develops, involving protonic and aprotic surface acidic sites that yield a donor-acceptor (acid-base) linkage with the acrylamide carbamide group. Note that the latter directs the monomer orientation towards the inorganic surface.

REFERENCES

1. Lipatov, Yuri, Sergeeva, Ludmila, *Adsorption of Polymers*. NewYork: Halsted Press, 1974.
2. *The Chemical Physics of Solid Surfaces and Heterogeneous Catalysts*, D. A. King and D. P. Woodruff, eds. Amsterdam: Elsevier, 1982-1984.
3. Syrkova O.V., Tsvetkov V.K., Pak V.N. Study of interaction of poly(hexamethylenesebacamide) with copper and magnezium silicates. Zh. Priklad. Khim. 1986; **59**: 1131-32 (In Russian).
4. Gutmann, Viktor, *The Donor-Acceptor Approach to Molecular Interactions*, New-York: Plenum, 1978.
5. *Kinetics and Mechanism of a Formation and Transformation of the Macromolecules (Kinetika i Mekhanizm Obrazovanija i Prevrashcheniya Makromolekul)*, Moscow: Nauka, 1968 (In Russian).
6. Gorbunova E.V., Deev Yu.S., Ryabov E.A. Mechanism of the polymerization of lactams in the presence of transition metal oxides. Plast Massy 1980; 17-19 (In Russian).
7. Glebov A.N., Vasil'ev P.A., Poselyagin S.V. Study of polymeric metallic complexes of polyacrilic and polymethacrylic acids with titanium and copper. Vysokomol. Soedin. Ser. B 1980; **22**: 805-8 (In Russian).
8. Basova R.V., Kristal'nyi E.V., Pakuro N. T., Arest-Yakubovich A.A. Polymerization of polar monomers in the presence of an organobarium initiator. Vysokomol. Soedin. Ser. B 1975; **17**: 263-4 (In Russian).
9. Kocherga J.J., Baran A.A. Effect of nonionic water-soluble polymers on the electrokinetic potential of colloidal particles. Teor. Eksp. Khim. 1976; **12**: 640-7 (In Russian).

10. Lipatov Yuri, Sergeeva, Ludmila, Adsorption of polymers from solutions. Adv. Colloid and Surface Sci. 1976; **6**: 1-93.
11. Fomkin A.A., Avramenko V.A., Seliverstova I.I., Serpinskii V.V. Adsorption of gases, vapor, and liquids in micropores as a unique phenomenon. Dokl. Akad. Nauk SSSR 1986; **288**: 678-81 (In Russian).
12. Lipatov, Yuri, *Interfacial Phenomena in Polymers* (*Mezhfaznije Yavleniya v Polimerakh*). Kiev: Naukova dumka, 1980 (In Russian).
13. Lipatov Yu.S. Role of interfacial phenomena in the formation of micro- and macroheterogeneities in multicomponent polymer systems. Pure and Appl. Chem. 1975; **43**: 273-89.
14. Lipatov Yu.S., Fabulyak F.G., Shifrin V.V. Effect of the chemical modification of filler surface on the behavior of a polymer in a boundary layer of varying thickness. Vysokomol. Soedin. Ser. A 1976; **18**: 763-6 (In Russian).
15. Zaporozhskaya E.A., Ginzburg L.V., Dontsov A.A., Romanov T.V., Fomin A.G. Effect of structure of phenol-formaldehyde resins on their reactions with α-aluminum oxide. Vysokomol. Soedin. Ser. A 1982; **24**: 1547-51 (In Russian).
16. Platonov B.E., Polishchuk T.A., Solov'eva T.E. Formation and structure of the interfacial layers of nonionogenic polymers in an interfacial boundary oxide-aqueous solution. Kompozits. Polymer. Materialy 1983; 69-72 (In Russian).
17. Filyanov E.M. Effect of filler on glass-transition temperature of an epoxy binder and its relation with the properties of a filled polymer. Vysokomol. Soedin. Ser. A 1978; **20**: 1845-8 (In Russian).
18. *Calorimetry and Thermal Analysis*, Gatta G. Della, ed. Amsterdam: Elsevier, 1985.
19. Kozlov S.G., Glushchenko V.Yu. Study by liquid chromatography of the temperature dependence of the adsorption of water-salt nitrophenol solutions on graphite. Kolloidn. Zh. 1976; **38**: 577-9 (In Russian).
20. Cosgrove T. *Polymers at Interfaces*. Chem. Ind. (London) 1988; 45-49.
21. Osmanov T.O., Gromov V.F., Khomikovskii P.M., Abkin A.D. Solution polymerization of acrylamide in the presence of Lewis acids. Vysokomol. Soedin. Ser. A 1980; **22**: 668-75 (In Russian).
22. Pshenitsyna V.P., Produvalova S.S., Molotkova N.N., Gorbunov N.N. Structure of melamine-formaldehyde oligomers. Vysokomol. Soedin. Ser. B 1980; **22**: 711-3 (In Russian).
23. Tolstaya S.N., Mikhailova S.S., Taubman A.B., Uvarov A.V. Adsorption of surface-active substances and polymers on titanium dioxide. Dokl. Akad. Nauk SSSR 1968; **178**: 148-51 (In Russian).
24. Kabanov D.A., Tkachenko A.G., Syrkova O.V., Tsvetkov V.K. Study of structure formation of composites based on copper sulfide-filled polyamides. Zh. Prikl. Khim. 1986; **59**: 1638-40 (In Russian).
25. Eremenko B.V., Platonov B.E., Baran A.A., Mamchenko A.V. Adsorption of poly(oxyethylene) and elektrokinetic potential of antimony sulfide particles in aqueous solutions. Kolloidn. Zh. 1975; **37**: 1083-9 (In Russian).
26. Lipatov, Yuri, Sergeeva, Ludmila. "Investigation of adsorption of polymers from solutions on glass fibers". In: *Ionoobmen i Sorbtsiya iz Rastvorov*, Minsk: Akad. Nauk Belorussk. SSR, Inst. Obshch. i Neorgan. Khim., 1963 (In Russian).
27. Masters, Christopher, *Homogeneous Transition Metal Catalysis*, London: Chapman and Hall, 1981.
28. *Molecular Interactions and Electronic Processes in Liquids* (*Molekul'arnie Vzaimodejstvia i Elektronnie Processy v Technostiah*), Novosibirsk: Nauka, 1987 (In Russian).

29. Votteler Ch., Hoffmann V., Trafara G. Modifications of polyamide-6 and polyamide-4 intercalated with alkali halides. Makromol. Chem. 1984; **185**: 1953-77.
30. Marmo M.J., Mostafa M.A., Jinnai H., Fowkes F.M., Manson J.A. Acid-base interaction in filler-matrix systems. Ind. Eng. Chem., Prod. Res. Dev. 1976; **15**: 206-11.
31. Rao K.H., Forssberg K.S.E., Forsling W. Interfacial interactions and mechanical properties of mineral filled polymer matrix. Coll. and Interf. A 1998; **133**: 107-17.
32. Heublein G., Hortschansky P., Klein S. Adsorption of flexible macromolecules at oxide interfaces, Acta polym. 1988; **39**: 117-21.
33. Finlayson M.F., Shah B.A. The influence of surface acidity and basicity on adhesion of poly(ethylene-co-acrylic acid) to aluminum. J. Adhes. Sci. Technol. 1990; **4**: 431-9.
34. Papirer E., Perrin J.-M., Nanse G., Fioux Ph. Adsorption of poly(methyl methacrilate) on an α-alumina: evidence of formation of surface carboxylate bonds. Eur. Polym. J. 1994; **30**: 985-91.
35. Bune E.V., Zhuravleva I.L., Sheinker A.P., Bogachev Yu.S., Teleshov E.N. Effect of intermolecular hydrogen bonding on polymerization of methyl metacrylate and acrylamide in solutions. Vysokomol. Soedin. Ser. A 1986; **28**: 1279-83 (In Russian).
36. Murakami Yo., Suzuki T., Takegami Yo. Anionic polymerization of acrilamide initiated with the disodium salt of poly(ethylene oxide). Polym. J. 1985; **17**: 855-61.
37. Fowkes F.M. Acid-base contributions to polymer-filler interactions. Rubber Chem. and Technol. 1984; **57**: 328-43.
38. Shustorovich E. Bond making and breaking on transition-metal surfaces: theoretical projections based on bond-order conservation. Surface Sci. 1986; **176**: L863-L872.
39. Pavlov S.A., Bruk M.A. Radiation-induced polymerization of monomers on the solid surfaces in systems with strong adsorption bonding. Vysokomol. Soedin. Ser. A 1987; **29**: 118-24 (In Russian).
40. Mund S.L., Bruk M.A., Abkin A.D. Some regularities of independent and combined radiation polymerization of monomers adsorbed on mineral templates. Vysokomol. Soedin. Ser. A 1976; **18**: 2631-8 (In Russian).
41. Pavlov S.A., Bruk M.A., Abkin A. D. Study of the molecular transport and polymerization of monomers in an adsorbed state. Adsorbtsiya i Adsorbenty, Kiev, 1981, 61-67 (In Russian).
42. Pavlova L.V., Pavlov A.L., Izjumnikov A.L., Bruk M.A. Influence of adsorption linkage on the reaction capability of monomers and radicals in copolymerization. Dokl. Akad. Nauk SSSR 1988; **301**: 142-6 (In Russian).
43. Olazar M., Bilbao J., Romero A., Aquavo A.T. Polymerization on solid catalysts. III. Incidence on the kinetic of the properties of the catalyst. Afinidad, 1987; **44**: 337-40.
44. Bryk, Mikhail, *Polymerization on the Solid Surface of Inorganic Substances (Polimerizatsiya na Tverdoi Poverkhnosti Neorganicheskikh Veshstestv*). Kiev: Naukova Dumka, 1981 (In Russian).
45. Polushkin V.A., Diner V.A., Lokutsievskii V.A., Khan I.G. Study of products of styrene polymerization initiated by kaolinite. Vysokomol. Soedin. Ser. B 1984; **26**: 420-5 (In Russian).
46. Litvinenko ML, Diner VA, Polushkin VA. Chemistry of unsaturated substances. Proceedings of National Conference dedicated in memory of A. M. Boutlerov (1828-1886); Kazan, 1986 (In Russian).
47. Polushkin V.A., Diner V.A., Zelener Z.F., Lokutsievskii V.A., Yakutin V.A. Polymerization of isoprene adsorbed on kaolinite from the gaseous phase, Vysokomol. Soedin. Ser. A 1984; **26**: 2098-103 (In Russian).
48. Daly G.M., Pithawalla Y.B., Yu Z., El-Shall M.S. Interaction of Zn^+ with isobutylene in the gas phase and in clusters. Metal ion induced cationic polymerization. Chem. Phys. Lett. 1995; **237**: 97-105.

49. Rudin V.N., Melikhov I.V., Bozhevol'nov V.E. Topochemical crystallization and some technological ideas for its use. Khim. Promishlennost' 1993; 381-8 (In Russian).
50. Osmanov T.O., Gromov V.F., Khomikovskii P.M., Abkin A.D. Vysokomol. Soedin. Ser. A 1979; **21**: 1766-73 (In Russian).
51. Savost'yanov V.S., Kritskaya D.A., Ponomarev A.N. Kinetics of radiation-induced liquid-phase graft polymerization of acrylamide on polyethylene. Vysokomol. Soedin. Ser. B 1985; **27**: 66-9 (In Russian).
52. Tanabe K., Yamagushi T., Akyama K., Mitoh A., Iwabushi K., Isogai K. The catalytic synthesis of acylated compounds by using solid super acid. Proceedings of the 8th Int. Congr. on Catalysis; 1984 July 2-6; Berlin (West), vol. V.
53. Ogenko V.M. Chemical physics of dispersed material surfaces. React. Kinet. Catal. Lett. 1993; **50**: 103-13.
54. Yamanaka T., Tanabe K. A representative parameter, $H_{0,\ max}$, of acid-base strength on solid metal-oxygen compounds. J. of Phys. Chem. 1976; **80**: 1723-7.
55. Kruba L.E., Amfiteatrova T.A., Kozlov L.V. Effect of pigments on the flow and curing of thermosetting acrylic composites. Lakokras. Mater. Ikh Primen. 1984; 14-16 (In Russian).
56. Tsuchida E., Yamamoto K., Nishide H., Yoshida S. Poly(p-phenylene sulfide)-yielding polymerization of diphenyl disulfide by S-S bond cleavage with a Lewis acid. Macromolecules 1987; **20**: 2030-1.
57. Aripov E.A., Ablyaev E.Sh., Vorob'ev V.N., Abdullaev N.F. Effect of activation temperature on concentration of Lewis and Brønsted acid centers of montmorillonite. Zh. Prikl. Khim. 1986; **59**: 1888-90 (In Russian).
58. Vaugeois J., Simard M., Wuest J.D. Multiple coordination of carbonyl compounds by multidentate Main group Lewis acids. Coord. Chem. Rev. 1995; **145**: 55-73.
59. Nel J.G., Wagener K.B., Boncella J.M. Acyclic diene metathesis polymerization. The synthesis of perfectly linear poly(1,4-butadiene). Polym. Prepr. (Am. Chem. Soc., Div. Polym. Chem.) 1989; **30**: (130-1.
60. Kotel'nikov V.A., Kurashev V.V., Danilevskaya L.B., Konova I.O., Gavrilenko V.V., Chekulaeva L.A., Garbunova I.A., Persits I.E. Active centers and some features of anionic polymerization of ε-caprolactam initiated with magnesium lactamates. Vysokomol. Soedin. Ser. A 1992; **34**: 69-76 (In Russian).
61. Bryk M.T., Varavko I.A., Kurilenko O.D. Thermal polycondensation of diphenylsilanediol containing dispersed titanium dioxide, iron (III) oxide, cobalt (III) oxide, and copper (II) oxide. Vysokomol. Soedin. Ser. A 1975; **17**: 2434-40 (In Russian).
62. Mikhael M.G., Padias A.B., Hall H.K., Jr. Effect of ZnCl2 on the spontaneous copolymerization of methyl acrylate with substituted 1,3-dienes. Acta Polym. 1996; **47**: 16-19.
63. Hutchinson R.A., Ray W.H. Polymerization of olefins through heterogeneous catalysis. VIII. Monomer sorption effects. J. Appl. Polym. Sci. 1990; **41**: 51-81.
64. Matvelashvili G.S., Gurman I.M., Potekhina E.S., Pukhovitskaya A.N., Smirnova L.N., Tarakhunov O.A. New polymer materials based on amino resins. Plast Massy 1980; 38-39 (In Russian).
65. Bessonov Yu.S., Tager A.A., Yushkova S.M., Komarov N.N., Guseev V.V., Rafikov M.N. Thermodynamic study of interactions of filled compositions based on poly(vynil chloride). Vysokomol. Soedin. Ser. A 1978; **20**: 99-105 (In Russian).
66. Tager A.A., Yushkova S.M., Bessonov Yu.S., Guzeev V.V., Rafikov M.N., Ezhov V.S. Thermodynamic study of interaction in filled plasticized poly(vinyl chloride) compositions. Vysokomol. Soedin. Ser. A 1979; **21**: 1051-8 (In Russian).
67. Tager A.A., Yushkova S.M., Safronov A.P. Interaction of fillers with polymers and their low-molecular-weight analogs. Plast Massy 1987; 26-7 (In Russian).

68. Klimenko N.A., Polyakov V.Ye., Permilovskaya A.A. Calorimetric study of adsorption of surfactants from aqueous solution on acetylene black. Kolloidn. Zh. 1979; **41**: 1081-6 (In Russian).
69. Little, Leslie, *Infrared Spectra of Adsorbed Species*, London: Acad.Press, 1966.
70. Naito Sh. Surface structure and catalytic activity. Hyomen Gijutsu 1995; **46**: 7-12.
71. Madix R.J. Surface reactivity and surface spectroscopies, Ultramicroscopy 1989; **31**: 58-66.
72. Chao Li Chung, Chang Eng-Pi Interaction of anhydrous ferric chloride with nylon 6. J. Appl. Polym. Sci. 1981; **26**: 603-10.
73. Pavlov S.A., Bruk M.A., Abkin A.D. Study of the molecular transport and polymerization of monomers in an adsorbed state. Adsorbtsiya i Adsorbenty Kiev 1981; 61-7 (In Russian).
74. Fukushima Yo. Montmorilonite/nylon hybrid compounds. Zeoraito 1991; **8**: 1-6.
75. Furukawa, Junji, Saegusa, Takeo, *Polymer Reviews. Vol. 3. Polymerization of Aldehydes and Oxides*, New York: Interscience, 1963.
76. Meijers S., Ponec V., Finocchio E., Busca G. FTIR study of the adsorption and transformation of allylbenzene over oxide catalysts. J. Chem. Soc., Faraday Transl. 1995; **91**: 1861-9.
77. Pokrovskii E.I., Novichkova L.M., Rostovskii E.N. Interaction on mineral dispersed phases with polymer functional groups. Dokl. Akad. Nauk SSSR 1970; **194**: 370-2 (In Russian).
78. Trubitsyna S.N., Lebedeva Yu.V., Askarov M.A. Emulsion polymerization of nitrogen-containing monomers. Izv. Vyssh. Uchebn. Zaved., Khim Khim. Tekhnol. 1976; **19**: 1429-32 (In Russian).
79. Laptii S.V., Kercha Yu.Yu., Gaiduk R.L., Vatulev V.N., Kosenko L.A., Shtompel V. I. Study of changes in the structure of segmented polyamide-polyether-polyurethane in the presence of lithium chloride. Vysokomol. Soedin. Ser. B 1985; **27**: 93-8 (In Russian).
80. Lipatova T.E., Sheinina L.S. On the effect of Aerosil on the kinetics of formation of linear polyurethanes. Vysokomol. Soedin. Ser. B 1976; **18**: 44-7 (In Russian).
81. Ionina N.V., Avstriiskaya E.E., Berezkina A.P., Nel'son K.V. IR-spectroskopic study of the molecular structure of highly filled compositions of low-molecular-weight diene rubbers. Vysokomol. Soedin. Ser. A 1978; **20**: 1124-9 (In Russian).
82. Shaboldin V.P., Demishev V.N., Ionov Yu.A., Akatova S.P. Effect of mineral salts on the structure of liquid Thiokol. Vysokomol. Soedin. Ser. A 1982; **24**: 1099-102 (In Russian).
83. Mitchenko Yu.I., Lebedeva T.I., Tsiperman R.F. Molecular interactions in polyamide-salt solutions. Vysokomol. Soedin. Ser. A 1984; **26**: 2542-7 (In Russian).
84. Trubitsyna S.N., Ismailov I., Askarov M.A. Role of ionogenic emulsifiers in low-temperature polymerization of nitrogen-containing monomers. Vysokomol. Soedin. Ser. A 1978; **20**: 1624-8 (In Russian).
85. Pshenitsyna V.P., Molotkova N.N., Smirnova L.N., Pukhovitskaya A.N., Chesnokova I.A. Curing of urea-formaldehyde oligomers. Plast Massy 1987; 20-2 (In Russian).
86. Davidov A.A., Loiko V.E. IR spectroscopic study of ethylene oxide polymerization on magnesia and calcia Zh. Prikl. Spektrosk. 1992; **56**: 209-15 (In Russian).
87. Gromov V.F., Khomikovskii P.M. Effect of a solvent on the rates of chain propargation and termination in radical polymerization. Usp. Khim. 1979; **48**: 1943-67 (In Russian).
88. Acierno D., Bianchi E., Ciferri A., De Cindio B., Migliaresi C., Nicolais L. Bulk properties of synthetic polymer-inorganic salt systems. III. Flow behavior and glass transition of salted polycaproamide. J. Polym. Sci., Polym. Symp. 1976; No. 54 (Polym. Sci.: Achiev. Prospects, Symp., 1975), 259-69.
89. Siegmann A., Baraam Z. Nylon-6 containing metal halides. I. Melting and crystallization. Int. J. Polym. Mater. 1980; **8**: 243-53.

90. *Ionomers: Characterization, Theory, and Applications*, Schlick Sh., ed. Boca Raton, Fla: CRC, 1996.
91. Judenstein P., Sanches, C. Hybrid organic-inorganic materials: a land of multidisciplinarity, J. Mater. Sci. 1996; **6**: 511-25.
92. Packham D.E. Engineering the interface in composites and bonded components. J. Adhes. 1995; **54**: 133-43.
93. Lee L.H. Molecular bonding mechanism for polymer and solid adhesion. Polym. Mater.Sci. Eng. 1993; **67**: 1-3.
94. Drago R.S., Fogel G.C., Needham T.E. Four-parameter equation for predicting enthalpies of adduct formation. J. Am. Chem. Soc. 1971; **93**: 6014-26.
95. Drago R.S., Parr L.B., Chamberlain C.S. Solvent effects and their relation to the E and C equation. J. Am. Chem. Soc. 1977; **99**: 3203-9.
96. Tanabe, Kozo, *Solid Acid and Bases:Their Catalytic Properties*, N. Y.: Academic, 1971.
97. *Acid-Base Interactions: Relevance to Adhesion Science and Technology*, K. L. Mittal and H. R. Anderson Jr., eds. Utrecht, The Netherlands, Tokyo, Japan, 1991.
98. *Surface Chemistry and Adsorption: Experimental Methods in Adsorption and Molecular Chromatography* (*Khimia Poverkhnosti i Adsorbtsiya: Eksperimental'nye Metody v Adsorbtsii i Molekularnoi Khromatografii*), Dreving V. P., ed. Moscow, USSR: Izd. Mosk. Univ., 1973.
99. Fukushima Y., Usuki A. Structure of montmorilonite/polyamide intercalated compounds. Funtai oyobi Funmatsu Yakin. 1992; **39**: (1045-9.
100. Jonathan N. The infrared and Raman-spectra and stucture of acrylamide. J. Molec. Spectroscopy 1961; **6**: 205-14.
101. La Mantia F.P., Acierno D. Glass transition behaviour of salted nylon 6. Coll. Polym. Sci. 1981; **259**: 693-6.

Chapter 3

RHEOLOGICAL BEHAVIOUR AND PROPERTIES OF FILLED POLYMER COMPOSITES

1. GENERAL PRINCIPLES

Reiner [1] defines rheology as a science that treats body deformation and flow. Founders of modern rheology consider deformation of mainly solid bodies, regarding them as homogeneous continuous media. The present study focuses on the rheological behaviour of filled polymers, where the filler affects not only material exploitation characteristics, but essentially determines its fabrication into a final product. A combination of the characteristics of the dispersed medium with those of the filler-formed structure is the basis of the rheology of filled polymers. Those systems may display different properties, depending on medium viscosity and strength of the structure skeleton of the filler. Modern rheology is dominated by two main problems; Malkin [2] defines them as follows:

- The first goal of rheology is to search for stress-deformation relationships for various technological and engineering materials in order to solve macroscopic problems related to continuum mechanics of these materials;
- The second problem consists in the establishment of relations between rheological properties of a material and its molecular composition. This goal is specific for microrheology, a trend founded by Reiner.

Note that not only temperature determines the rheological behaviour of polymers, polymer solutions, and polymer-based filled compositions. Polymer nature, molecular weight, and molecular weight distribution (MWD), together with flow-initiating shear stress and strain, also affect the rheology of polymer compositions. To characterize polymer rheology and to formulate flow criteria, one needs to establish *viscosity-shear stress* or *shear stress-shear strain rate* relations. The simplest case is when polymer shear stress is proportional to the shear strain rate:

$$\tau = \eta . \dot{\gamma} \tag{3.1}$$

This is in fact Newton's law, where τ is the shear stress in N/m^2, $\dot{\gamma}$ is the gradient of shear strain rate in s^{-1}, and η is the coefficient of proportionality. In literature the characteristic η is known as a "viscosity coefficient" or "viscosity", with dimensions Pa.s.

Curve 1 in Fig. 1 is the simplest flow curve, typical for Newtonian fluids. In practice, Newtonian behaviour is confined to liquids with low molecular weight. However, technologists seldom consider such type of behaviour during polymer processing. Polymer melts obey Newton's law only for a shear rate ($\dot{\gamma}$) close to zero. Moreover, polymer solutions and composites obey Newton's law for concentration close to zero. Before considering the rheology of polymer composites, we shall describe various types of rheological behaviour as based on the $\tau/\dot{\gamma}$ curve type (Fig. 1). As is seen in Fig. 1, the flow curves significantly differ from one another. For a *pseudoplastic flow*, for instance, (curve 3 in Fig. 1) viscosity is not constant but decreases with the increase of the shear rate. For *dilatant fluids* (curve 2 in Fig. 1) viscosity increases with the rise of $\dot{\gamma}$. Note that the shear stress for pseudoplastic liquids increases less than proportionally with the increase of the shear rate while for dilatant liquids the increase is larger than proportional. An equation known in rheology as *power law* describes these two liquids:

$$\tau = \eta . \dot{\gamma}^n \tag{3.2}$$

where η is the Newtonian viscosity (i.e. viscosity for $\dot{\gamma} = 0$). The exponent n may take the following values: $n > 1$ for dilatant liquids and $n < 1$ for pseudoplastic liquids. A Newtonian liquid, $n = 1$, is a special case.

Bingham behaviour is typical for higher concentrations of polymer solutions and compositions (curve 4 in Fig. 1). The following equation describes such type of behaviour:

$$\tau - \tau_0 = \eta . \dot{\gamma} \tag{3.3}$$

where τ_o is the yield stress, (i.e. *yield value* or *yield point*). This is the limit stress over which flow initiation takes place. After the flow start, viscosity depends strongly on $\dot{\gamma}$, i.e. $\eta = f(\dot{\gamma})$. Bingham observed such behaviour when studying a clay dispersed system; it is typical for greases, fats, toothpaste, cement paste, drilling mud, sewege sludge, etc.

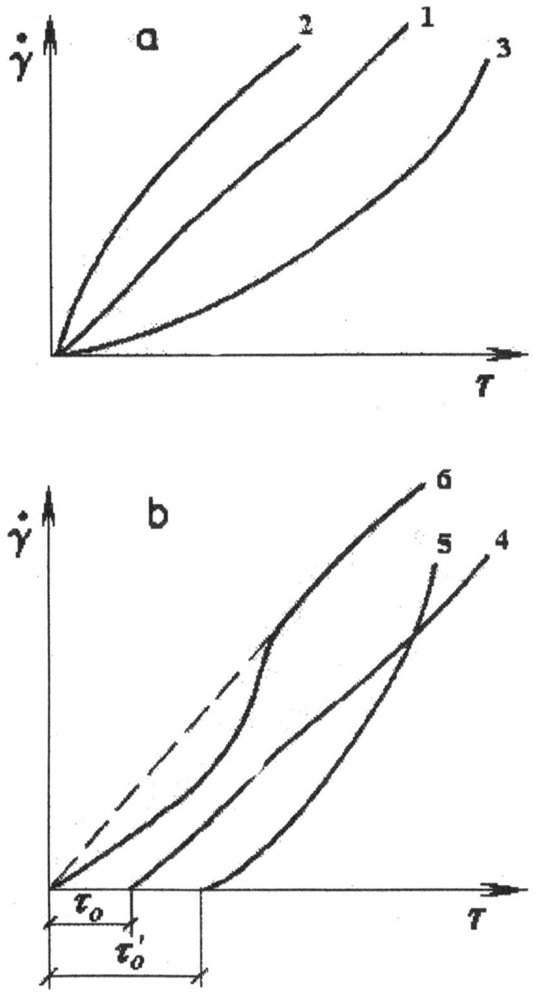

Figure 1.

Curve 5 in Fig. 1 shows that the rheology curve of plastic flow is curvilinear and differs from the Bingham curve. Another type of rheological behaviour is that of Ostwald (curve 6). Rehbinder described it by means of a full rheology curve [3]; two different values of Newtonian viscosity characterize this rheological behaviour (Fig. 2.). The first *maximal Newtonian viscosity* (η_0), is characteristic for small values of $\dot{\gamma}$; here the flow of the polymer system and the destruction of its structure have not practically started ($\tau < \tau_o$). The second *minimal Newtonian viscosity* (η_m) is attained for large values of $\dot{\gamma}$, where the destruction of the polymer system is total. Viscosity in this case is constant, depending not on $\dot{\gamma}$, but on the characteristics of the filled polymer. The intermediate section of the rheology curve corresponds to a viscosity which changes as a function of the applied $\dot{\gamma}$ or τ. Such dependence holds for a number of real cases, where viscosity is known as *apparent* (non-Newtonian). Some researchers call it *structural viscosity*, since its change reflects the change of the structure during the flow. It was Ostwald who first outlined viscosity change for colloid dispersions. He explained the phenomenon as a change of the colloid structure due to deformation. The apparent viscosity decreases; the effect is known as *shear thinning*. Note that viscosity may also increase with the increase of the shear rate - a phenomenon known as *shear thickening*. Finally, viscosity of a given material may decrease or increase for different ranges of the shear rate.

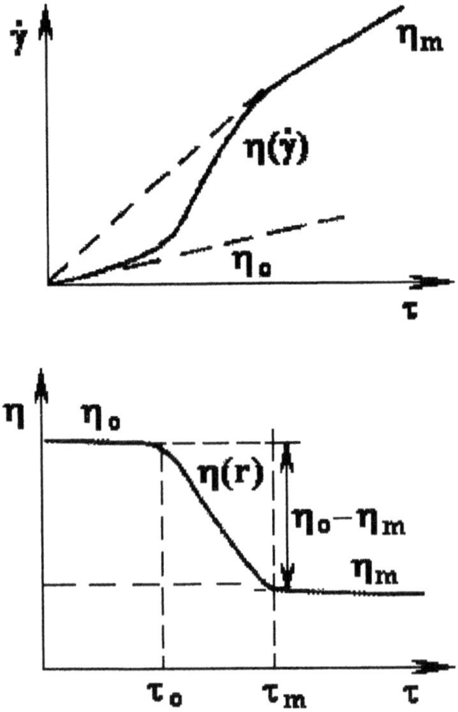

Figure 2. Flow curve of non-Newtonian fluid

A polymer fluid manifests *pseudoplastic* behaviour when: (i) its flow is characterized by a growing shear rate; (ii) the flow curve curvature is convex with respect to the shear rate axis. For very low values of $\dot{\gamma}$ the curve slope is constant, and there is a critical value of $\dot{\gamma}$ which marks the start of deviation from linearity. Below the critical value of $\dot{\gamma}$, the slope has the maximal value of the Newtonian viscosity, known in the literature as *zero shear viscosity*.

Above the critical value of $\dot{\gamma}$, the polymer is considered to be *shear-softened*, in full accordance with the observed strain softening of hard polymers. Polymer melts and polymer solutions filled with mineral filler display such a rheological behaviour; they posses not only maximal, but also minimal Newtonian viscosity.

For dilatant rheological behaviour, the curve is curved in the opposite way, concave with respect to the $\dot{\gamma}$ axis. Some higly concentrated filled polymer systems posses such a rheological behaviour. Bingham rheological behaviour is not so common for polymers, but is often characteristic for various highly filled systems. The yield stress of Bingham systems marks flow initiation and viscosity irregularity, i.e viscosity decrease with yield stress increase. Regarding the plastic behaviour of such systems Lenk [4] outlines that "it is by no means certain that this plastic type does in fact exist". If it does exist, then the generalized flow curve can readily account for it on the assumption that the first (maximal) Newtonian viscosity coincides with the shear stress coordinate. Such an assumption is equivalent to the statement that the system is characterized by an initial infinite Newtonian viscosity, since the low shear portion of the curve coincides with the shear stress coordinate up to a point where a strain *yield stress* is reached. This means that flow, similar to the pseudoplastic one, starts after attaining a definite yield stress.

As noted above, polymer system viscosity depends on a number of factors and the dependence generally reads:

$$\eta = f(\dot{\gamma}, T, t, P, \phi, i, \ldots) \quad (3.4)$$

Here η is viscosity, $\dot{\gamma}$ - shear rate (itself a function of the shear stress), T - temperature, t - time, P - pressure, ϕ - concentration, i- a parameter, characterizing the interfacial phenomena and processes. Dots stand for polymer molecular weight, molecular weight distribution, the composition variable (crystallinity) and additives (plasticisers, fillers, pigments, stabilizers, impurities, etc.).

A number of authors use different analytical functions to describe the flow curve of various materials both theoretically and empirically. The most popular forms of the flow curves adopt an expressed value of the initial (maximal) Newtonian viscosity. They read as follows:

Cross [5]:

$$\eta = \frac{\eta_0}{1+[\lambda\dot{\gamma}]^m} \quad (3.5)$$

Carreau [6]:

$$\eta = \frac{\eta_0}{\left[1+[\lambda\dot{\gamma}]^2\right]^{-p}} \tag{3.6}$$

Yasuda et al. [7]:

$$\eta = \frac{\eta_0}{\left[1+[\lambda\dot{\gamma}]^a\right]^{(n-1)/a}} \tag{3.7}$$

Vinogradov-Malkin [2]:

$$\eta = \frac{\eta_0}{1+A\dot{\gamma}^a + B^{2a}} \tag{3.8}$$

Schulman [8]:

$$\tau^{1/n} = \tau_0{}^{1/n} + (\dot{\gamma}\,\eta)^{1/m} \tag{3.9}.$$

The quantity η in all the equations is the apparent viscosity that depends on the shear rate. The parameter η_0 is the zero-shear viscosity and the remaining parameters are empirical or semi-empirical constants. The equations are similar in structure but differ from one another with respect to the number of parameters involved. Note, however, that all include a limit η_0 for $\dot{\gamma} \rightarrow 0$ and a limit zero for $\dot{\gamma} \rightarrow \infty$.

Van Oene [9] attempts to estimate flow of polymer blends and suspensions. Besides, he notes that polymer blends may be treated as dispersions of deformable liquid particles, while reinforced composites may be treated as dispersions of axisymmetric solid particles. A number of factors determine the rheology of suspensions: dispersion structure, shape and orientation of the dispersed particles, particle interaction and interfacial phenomena.

Real materials (including polymer filled composites) change their properties very often in time, due to structural transformations which arise from the action of applied shear, and rebuilding their initial structure at rest. The change of properties may be fully reversible – a phenomenon called *thixotropy* [2, 4]. This material time-dependent behaviour is defined as isothermal and reversible time dependent loss of viscosity arising from the applied shear. The process of structure rebuilding is often extremely slow. To accelerate this process the material may require special treatment, such as temperature increase.

Recently, there has been growing interest in interfacial chemical reactions that take place in polymer blends. Blend chemistry is an object of theoretical and applied study, since it is quite important for the blend compatibilization and for the filled systems. As a matter of fact, one may generate block or/and graft copolymers or even a crosslinked interface, depending on the polymer nature and functionality. Hence, one may attain particle size reduction and/or morphology stabilization through a coalescence reduction. Cirman and co-workers [10] study the interface formed in a reactive blend by performing bulk viscoelastic measurements.

2. RHEOLOGY OF FILLED POLYMER SYSTEMS

Lets now focus our attention on the interfacial interactions that take place in filled polymer systems. We shall assess how these interactions affect the rheology and properties of the polymer systems. Note that a number of papers have been recently published in the last 7-10 years [11-22]. They regard important problems of polymer filling – filling degree, filler type, filler particle size and shape, etc. Note also the significant number of monographs on filled polymer systems – [3, 23-29], as well as the detailed reviews - [30-32].

The analysis of literature on the rheology of filled polymers and elastomers reveals that most authors deal with polymers whose filler content per unit volume is low. Few publications discuss the rheology of polymers of high filler content which has a definite influence on the degree of interfacial interactions. [30] presents a detailed analysis of the rheology of highly filled polymer melts; [22, 28, 33] deal with highly filled polymer composites.

One may explain the recent interest in highly filled polymer systems by the widespread introduction of composites with a polymer matrix or with a polymer as binder. Based on empirical models, literature provides profound analysis of the relation $\eta = f(\phi)$; here ϕ is a filler content. Consider concentrated liquid polymer compositions. Then, the coefficient ϕ_{max} (maximal volume filling) gives a better description of the effect of the filler on viscosity. Roscoe [33] and Maron & Pierse [34] propose this idea, claiming that a *concentration normalization* takes place. Table 1 presents the most popular empirical formulas for relative viscosity $\eta_r = f(\phi)$.

Surveying the rheology data on filled systems, one may establish that the majority of works tackle a narrow field of study. The publications do not allow for the formulation of a number of general laws that should control the behaviour of these compounds.

Table 1. Empirical formulas for $\eta_r = f(\phi)$ for concentrated liquid composites

Authors	Empirical formula
Chong J., Bear A., Cristiansen E. [35]	$\left(1+0.75+\frac{\phi/\phi_m}{1-\phi/\phi_m}\right)^2$
Lee D. [36]	$(1-\phi)^{-\left(2.5+1.92\phi+7.739\phi^2\right)}$
Brady J., Vicic M. [37]	$\frac{1.3}{(1-\phi/\phi_m)^2}$
Ivanov Ya., Kotsilkova R. [38]	$exp\left(\frac{1}{\alpha - \beta\phi}\right)$
Ghosh A., Bhattacharya S. [39]	$\frac{\left[1-c(\phi/\phi_m)\right]^2}{1-\phi/\phi_m}$
Eilers H. [40]	$\left(1+\frac{[\eta]\phi}{2(1-\phi/\phi_m)}\right)^2$
Graham A., Steel R., Bird R. [41]	$\left\{1-\phi\left[1+\left(\frac{1-\phi}{\phi_m}\right)\sqrt{1-\left(\frac{\phi_m-\phi}{\phi_m}\right)^2}\right]^2\right\}$
Ham R. [42]	$(1-\phi/\phi_m)^{-1}$

Authors	Empirical formula
Tomas D. [42]	$1+2.5\phi+10.05\phi^2+A\,exp(B\phi)$
Gillespie Th. [43]	$\frac{1+\phi/2}{(1-\phi)^2}$
Barnea E., Mizrahi J. [44]	$exp\frac{5}{3}\left(\frac{\phi}{1-\phi}\right)$
Leighton O. [45]	$\left(1+\frac{1.5\,\phi}{1-\phi/\phi_m}\right)^2$

Now consider liquid polymer composites with various fillers added. As a number of authors find, the compounds display quite a complex rheology and its general description is a difficult task. One should consider several factors that act simultaneously and reflect the features of (a) the polymer matrix, (b) the structure and structure formation of the dispersed filler and (c) the matrix-filler interaction. Note also that each of these factors depends, in different ways, on temperature, deformation type, stress magnitude, etc. In addition, composite morphology should be considered when analyzing the compound properties. As in filled systems, the first postulate of structureless filler and polymer requires that both phases lack micro defects. The second postulate states that there is an ideal adhesive polymer-filler contact; this follows from the condition for retaining both the continuity of the filled system and that of the polymer-filler interface, when an external field acts on it. Practically, an optimal adhesive contact between two phases is possible provided that the energy of interaction at the phase interface is higher than the energy of cohesion of one of the phases. A combination of a *low-energy* continuous phase (polymer) and a *high-energy* dispersed phase (minerals, fibers, etc.) satisfies this condition.

As we have said, the filler type and its acid-base surface properties are important for the polymer structure formation. A question arises as to how to choose the appropriate filler when preparing filled polymer compositions. Filler importance grows significantly when we consider its active participation in the acid-base interactions - see Chapter 1. A possible interaction between the filler and the polymer matrix is an essential aspect of the overall behaviour of filled polymer systems.

We note that the rheology characteristics of filled basic polymers vary within definite ranges as a result of filler acid activity. Carbon black and fumed silica are two of the industrially important reinforcing fillers. A lot of work has been concerned with how these two fillers affect the rheology of polymer melt suspensions [23, 46, 47]. However, one of the significant difficulties encountered when interpreting rheology data, is how to determine the effective volume fraction of the filler. Since carbon black and fumed silica have very complicated structure, one must be concerned with the effect of occluded and bound rubber which increases the effective filler volume. Some research [48, 49] has been done on systems with model filler particles. However, the problem is that those particles are fairly large - > 1 μm. As known, the bound rubber effect for such particle size is more or less obscure. De Groot & Macosco [50] consider systems that consist of linear poly(dimethylsiloxane) (PDMS) and model silica particles. The silica particles have an average radius of 39 nm. The polymer and filler are mixed in solution to prevent aggregation.The study indicates that a physical network of polymer-filler particle bonds is generated above a critical loading level. The scaling of the dynamic moduli in the linear viscoelastic region of the network seems to be similar to that for a polymer network, covalently bounded (unfilled) above the gel conversion. One should ask whether suspensions of more complex particles (for instance, fumed silica) could display a similar behaviour. Data on fumed silica [47, 50] seem to confirm this supposition. Recent experimental evidence indicates that ageing significantly affects the modulus of suspension prepared by using the solution method. This may result from the slow rate of PDMS adsorption on silica. The time of attaining to equilibrium may be too large [51]. The consequences of this ageing effect on the validity of time - filler level superposition are still obscure and need further analysis.

One may consider agglomeration that takes place during polymer-fumed silica interaction and which determines different types of interactions. Such are: (a) direct bridging; (b) bridging through entanglement of nonadsorbed chains and (c) bridging through entanglement of adsorbed chains. Considering polymers of low and high molecular weight [63] notes that their bonding to the silica surface is different. This will affect the rheology of the overall system.

Marti-Martinez et al. [51] study the properties of solvent based polyurethane adhesives, containing fumed silicas. They analyze compositions with five hydrophilic and two hydrophobic fumed silicas having different surface area and particle size. They establish a noticeable increase of the adhesive viscosity, imparted negative thixotropy and increased storage modulus G'. They also observe an improved adhesion of the chlorinated rubber/PU adhesive/chlorinated rubber joint. Those modifications are more pronounced in adhesives which contain hydrophilic silicas.

Composites, in general, and polymers filled with silica particles, in particular, are interesting for practice and theory [52]. The properties of both phases and, to a large extent, the interfacial charateristics determine the use of these compounds. [53] presents an interesting comparison of the rheology of filled polystyrene (PS) melts. The study considers also the ineraction between macromolecules and the filler particles. The matrix material is PS 100 filled with silica particles of different type (Degussa and Vacker, Germany) – Table 2.

Table 2. Characteristics of the used silica [53]

Type	S, m^2/g	Primary diameter d_p, μm	Diameter of aggregates D_A, μm	ϕ_{max}, %
R972	130	0.016	0.2	9
R974	200	0.012	0.2	7
HDK H30	300	0.07	0.2	5
HDK H20*	200	0.012	0.2	5

The study of REM-pictures of silica (R972) filled PS (1 wt. %) indicates some specific features of the compound: (i) formation of 200 nm aggregates; (ii) immobilization of the matrix around the aggregate; (iii) disentaglement in the middle between the aggregates, only. All this yields a nest formation.

Measurements of the composite rheology show a pronounced equilibrium modulus G_e at small frequencies and for any filling degree. When the value of G_e exceeds that of the plateau modulus G_p, the composite attains maximum volume concentration (ϕ_{max}). Note that the dependence of the equilibrium modulus on ϕ is much stronger than that of the plateau modulus. The study [53] follows the matrix-filler interaction, considering the change of the equilibrium modulus as an example. This study proposes a phenomenological model that includes an additional characteristic time λ_i and power law exponents a and b which depend on the polymer-filler interactions:

$$\frac{G' - G_e}{G_p - G_e} = \frac{(\omega\lambda_i)^b \left[cos\left(\frac{b}{2}\pi\right) + (\omega\lambda_i)^a cos\left(\frac{b-a}{\pi}\right) \right]}{1 + 2(\omega\lambda_i)^a cos\left(\frac{a}{2}\pi\right) + (\omega\lambda_i)^{2a}} \tag{3.10}$$

In most practical cases, the introduction of mineral particles accompanies the application of elastomers. This yields an increase of both the elastic modulus and fracture resistance. Although filled elastomers have found an application for years, there is still lack of a profound knowledge on their reinforcement mechanics. Clement and co-workers [54] perform an interesting study of a filled composition of poly(dimethylsiloxane) (PDMS) and pyro-silica. They mix silica with the poly(dimethylsiloxane), but before reticulation. In the structure so formed the polymer chains bond with each other and adsorb on the particle surface. This adsorption yields the formation of bonds between the acid silanol groups on the surface of the silica particles and the oxygen atoms of the polymer chains. As expected, this affects not only the mechanics of compound reinforcement, but also the compound rheological properties. The filled polymer displays a specific viscoelastic behaviour. It is expressed in a significant decrease of the G' modulus (from G'_0 to G'_∞) with the increase of the deformation amplitude. At the same time, loss modulus G'' attains a maximal value. Hence, energy dissipation, related to a change of the system structure, takes place. This behaviour is known as the *Payne effect* and depends on several factors - filler amount, surface acid properties of the silica particles and temperature, thus highlighting the importance of the interface between the filler and polymer. The authors of [54] describe it by employing the model proposed by Maier & Goritz, which assumes that there are two types of filler-matrix bonds - stable and unstable ones; the latter may easily break under external mechanical impact or temperature evolution. The bond formation leads to an increase in the density of system reticulation; this density has the following form:

$$V = V_c + V_{st} + V_i \tag{3.11}$$

where V_c is density of the chemical network reticulation, while V_{st} and V_i are the numbers of stable and unstable *matrix-particle* bonds per unit volume, respectively. Hence,

$$G'(\gamma) = [V_c + V_{st} + V_i]kT = G'_{st} + G'_i(\gamma); \tag{3.12}$$

where $G'_{st} = [V_c + V_{st}]kT$.

As in Langmuir's adsorption theory, the authors of [54] propose formulas expressing the dependence of the storage and loss moduli on $\dot{\gamma}$:

$$G'(\bar{\gamma}) = G'_{st} + G'_i \frac{1}{1+\phi\bar{\gamma}},$$

$$G''(\bar{\gamma}) = G''_{st} + G''_i \frac{\phi\dot{\gamma}}{(1+\phi\dot{\gamma})^2} \tag{3.13}$$

Here G'_{st} is the value of G' for maximal possible deformation. This means that when $\dot{\gamma}$ increases, modulus G' tends to G'_{st} - a value related to G'_{∞}. The modulus G'_o is a sum of the moduli G'_{st} and G'_i.

Note that the authors consider a system filled with 40 % silica. Depending on temperature, they obtain a very good agreement between the calculated and the experimentally obtained data for G' and G''. As expected, the modulus G'_i is more sensitive than G'_{st} to the temperature change. We should also note that the proposed model does not account for the filler degree of dispersity. Yet, based on simple physics, it elucidates the part of interfacial phenomena in the development of dissipative processes; it proposes an original molecular interpretation of the reinforcement.

The authors of [54] give data that illustrate the effect of the silica surface treatment. The modulus G' of a PDMS filled with treated silica strongly decreases. One should link this phenomenon with the mechanism of matrix-filler interaction. It consists in poisoning of silica particles and decrease of the number of active silanols.

[55] outlines the importance of silica modification by using a hydrophobic dispersion phase of octadecyl groups whose surface coverage is 2.6 $\mu mol/m^2$. The authors show that the surface modification strongly affects flow behaviour, especially at low shear rates. Suspensions of surface-modified particles display a significant increase of the yield stress – by more than one order. The model compounds consist of silicon oil and porous silica in original and surface-modified forms. In that case the stress increase is due to the intensification of inter-particle interactions. Comparison of the complex moduli or the values of $tan\delta$ shows an increase of the elastic component for the surface-modified suspensions. The authors explain this evidence by the brush-like elastic structure of the surface modified molecules, affecting particle interaction.

Chemisorbed and physically sorbed silanes play important roles in composite processing. The study [56] analyzes how the coupling agent affects the rheology of a model composite. The authors aim at the quantitative and qualitative assessment of the interface modification. Particle interactions and friction in poly(ethylene)/silicon dioxide depend on the acid surface coverage with the silane coupling agent. Hence, low silane coverage of the silica reduces the value of the dynamic shear modulus and $tan\ \delta$. High coverage, however, forms an interlayer at the interface. The physically adsorbed silane acts as a lubricant during composite processing [56].

Silane treatment of the filler increases its tensile strength by 1.5 – 2 times over the tensile strength of the matrix; it also diminishes the ultimate elongation. The treatment lowers melt viscosity and decreases the difference between the normal stresses for the applied identical shear stress [23, 57].

Particle-medium interactions depend both on particle surface and dispersed medium. Rheological properties of disperse systems are most sensitive to the interaction character. Hence, one often [3, 58] employs rheological properties as an integral characteristic for the assessment of particle-matrix interactions. It seems quite useful to consider the rheology of a polymer disperse system and that of a solid phase with particles having similar dimensions but different surface properties; [59] employs such approach. The authors use poly(methylsiloxan) PMS-1000 as a dispersing medium. They introduce aerosil A-300 with a specific surface S = 300 m^2/g, pyroalumina with S = 140 m^2/g and methylized aerosil MAS-200 with S = 200 m^2/g as highly dispersed solid phases. The natures of the surfaces of the compounds are different. As noted in Chapter 1, the aerosil display an acidic character but alumina is amphoteric, while the methylized aerosil should display basic character. One may perform rheological measurements and determine the dissipation of the mechanical energy for those three suspensions. The change of energy dissipation during the rheological measurements reads:

$$W = \int_0^t P.\dot{\gamma}.dt \tag{3.14}$$

Here P is the shear stress and t – time. Hence, the instantaneous rate of energy dissipation per unit dispersion volume has the form

$$w = P\dot{\gamma}. \tag{3.15}$$

The experimental evidence (Fig. 3) shows that phase interactions do not take place in a system with 1 % aerosil, but partial or entire disintegration of the primary aggregates occurs. We consider multi-level aggregates. Note that the flame reaction produces highly dispersed protonized particles ($d \approx 10$ nm); these particles form the aggregates at high temperature, and collide with each other.

Fig. 4 shows that the relations $\eta = f(\dot{\gamma}, \omega)$ significantly differ from each other for one and the same temperature at arbitrary low values of ω. For an increase of ω, however, relations $\eta = f(\omega)$ tend to equal values of η. We should note that some increase of η with a rise of the temperature at the definite value of $\dot{\gamma}$ is observed [59]. The authors show that regardless of the character of the dispersed particle surface, the dispersions have a yield stress τ_0, i.e. they are solid-like gels. However, one may better establish the structure difference by analyzing the value of τ_0. The authors obtain maximal values of τ_0 for A-300/PMS-1000 and minimal values of τ_0 - for MAS-200/PMS-1000. The values of τ_0 are intermediate for dispersions with γ-Al_2O_3. As seen, the quantity τ_0 takes different values. This is due to the following reasons. The SiO_2 particles, which have silanol groups on their surface, in the poly(methylsilane) interact more strongly, than with the methylized dispersed particles of MAS-200. The dispersions of MAS-200 and γ-Al_2O_3 in PMS behave as systems of weaker particle interaction, due to the strongly developed coatings. Hence, immobilization of significant portions of PMS takes place within the coatings. As a result the volume content of the dispersed phase significantly increases; this is accompanied by significant increase in the relative viscosity even for a 10 % volume fraction.

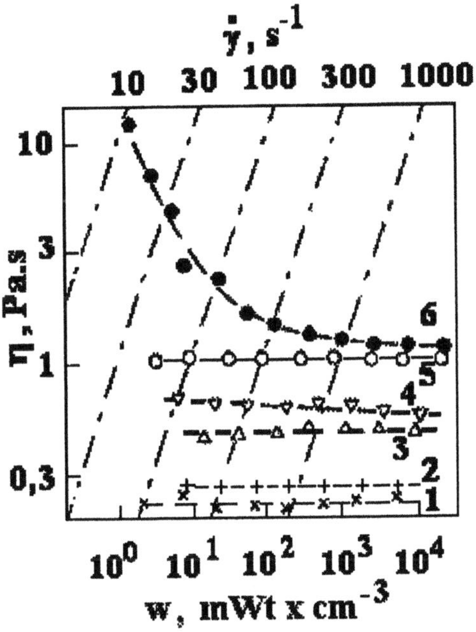

Figure 3. Effective viscosity of the PMS (*1, 3, 5*) and 1 %-dispersion of the aerosil in PMS (*2, 4, 6*) *vs.* dissipation rate of the mechanical energy (*W*) and the shear rate gradient ($\dot{\gamma}$): *1, 2* - 80 °C; *3, 4* - 60 °C; *5, 6* - 20 °C

Recently, researchers have paid much attention to the study of dispersed structures in various organic liquids [55, 60, 61]. This is because industry needs new electro-insulating coatings, plastic lubricants, varnish, paints, hermetics and other composites. Koshevar's study [62] in the field is of real interest; it is part of a systematic search for a correlation between the system macrocharacteristics – stability and specific interfacial interactions. The author analyzes disperse systems of zinc oxide–poly(methylmethacrylate)-solvent. It is worth noting that in this case both ZnO and PMMA are basic in caracter while the solvents have different behaviour: chloroform (Ch) and benzyl alcohol (BA) are acids, but cyclohexanon (CH) and ethylacetate (EA) are bases. Manufacturers use zinc oxide (ZnO) as filler and pigment for a number of industrial composites. Zinc oxide [62] has a specific surface of 17 m^2/g and average particle diameter of 5 μm. The molecular weight of the poly(methylmethacrylate) is 2.5×10^5. Koshevar found that the sedimentation rates are substantially different for the different media. For instance, the sedimentation rate of ZnO in an acidic medium (chloroform, benzyl alcohol), decreases due to a strong acid-base interaction with the increase of its volume fraction. The formation of a structural network may explain this phenomenon. The network is able to block separate particles or aggregates, thus stabilizing the system. The comparison indicates that the sedimentation rate in chloroform proves to be more stable than that in benzyl alcohol.

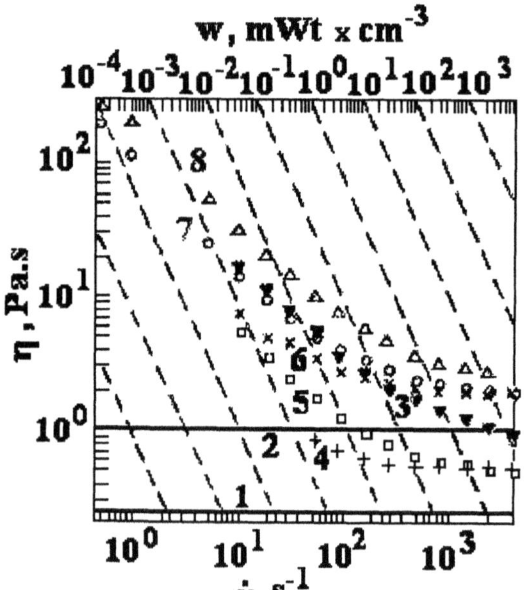

Figure 4. Effective viscosity of 5 % dispersion of aerosil, methylized aerosil, γ-Al_2O_3 in PMS and pure PMS *vs.* shear rate gradient ($\dot{\gamma}$) and dissipation rate of the mechanical energy (*W*): ***1*** - PMS, 80 °C; ***2*** - PMS, 20 °C; ***3*** - MAC, 80 °C; ***4*** - γ-Al_2O_3, 80 °C; ***5*** - MAC, 20 °C; ***6*** - γ-Al_2O_3, 20 °C; ***7*** - A-300, 20 °C; ***8*** - A-300, 80 °C.

Dispersions of a basic medium display opposite behaviour: the sedimentation rate increases with the ZnO content. This shows that the structure network is unable to block the particles. Owing to diffusion, they approach each other and aggregate. Acid-base interactions may explain this phenomenon, since both medium and phase are basic. As outlined in [62], *base-base* interaction is significantly weaker than *acid-base* one.

The study of the rheology of disperse systems with increasing degree of filling discloses their stabilization mechanisms. Compared to diluted systems, highly filled polymer systems attain stability in a different manner – see Table 3. Table 3 shows that viscosity η_{max} increases 100 times in chloroform when of the zinc oxide grows from 1.8 to 3.5 wt. %. The quantity η_{max} increases 35 times in benzyl alcohol and only 2 times - in cyclohexanon. The sedimentation of polymer-containing dispersions of ZnO particles is unstable in ethylacetate, which bars the increase of ϕ over 3.5 wt. %. The quantity η_{max} increases only 1.1 times in ethylacetate. The results show that the structure formation in acidic media takes place much more effectively than in basic media. Data on the yield stress τ_0 confirm this conclusion; Fig. 5 shows that the shear stress increases with ϕ for all dispersions, except for the system PMMA-ethylacetate-ZnO. The increase of ϕ is especially intensive for dispersions of chloroform and benzyl alcohol. Thus, the shear stress τ_0 of ZnO dispersions in solutions of PMMA, in chloroform and in cyclohexanon takes values of 5.5, 10 and 0.4 Pa.

However, if $\phi > 3.5$ %, the change of the structure strength of PMMA-chloroform-ZnO is more pronounced than that of a system with benzyl alcohol. [62] finds that for $\phi < 4$ % and the selected deformation of PMMA-chloroform-ZnO does not depart from Newtonian flow. For $\phi > \phi_{crit}$, however, phase separation occurs, and τ_0 cannot be determined. The data in Fig. 5 yields:

- sedimentation stability of polymer-ZnO dispersions decreases in basic medium with the increase of ϕ;
- this is because of the formation of a structure network that is too weak;
- this network either fractures under a moderate load, or lacks in the system PMMA-ethylacetate-ZnO.

The data of Fig. 5 show that when the ZnO concentration exceeds the critical value for forming a thixotropic structure, the shear stress τ suddenly increases in chloroform. Benzyl alcohol dispersions of ZnO have higher fluidity ($1/\eta_{max}$) than chloroform dispersions. This property allows for higher filling of the system and for smoother increase of τ_0.

Reconsider the experimental evidence for a highly filled system of PMMA-ZnO-solvent. One may link sedimentation/aggregation stability and rheology, and on the other hand, one may relate sedimentation/aggregation stability to the donor-acceptor properties of the medium. However, the link arises because of the different mechanisms of interaction at the various interfaces: particle-solvent, polymer-solvent and particle-polymer. One needs additional analysis to explain the nature of the link and to find the quantitative relations. Data prove that an account of the specific interfacial interactions is most important for the control of the structural-mechanical properties of the system.

[63] studies the kinetic stability and rheological behaviour of gel-like suspensions of talc-polyacrylglycinamide (PAG); it employs PAG polymer with branching structure, capable of forming thermo-reversible gel.

The curve of PAG adsorption on talc is characterized by the sharp increase at the begining of the equilibrium concentration and then a passage through a plateau (6.4 mg/g or 1.7 mg/m^2)- [63]. This course of the adsorption curve shows that there is strong affinity between PAG chains and the talc surface. Hence, both groups C = O and NH_2 may interact with protons on the surface. This is due to the existence of a majority of hydroxyl groups on its layer brucite although an adsorption of certain PAG segments on the basic hydrophobic surface of talc may occur. Moreover, water, being PAG solvent, may intensify PAG adsorption on talc. The lack of phase separation and talc sedimentation in specimens of test 1 (Table 4), and particularly in those of test 4, shows that PAG network aids particle stabilization. Considering PAG concentration (4 wt. %), we may conclude that talc particles (10 wt. %) are well dispersed in the suspension. Moreover, we can suppose that the segments of the adsorbed chains participate in the network formation and are caught around the particles during gel cooling and ageing [63]. This polymer network consists of branched PAG chains.

Table 3. Rheological properties of ZnO dispersions in various non-water media

Dispersion	ϕ, wt. %	η^{**}_{max}, Pa.s	η^{**}_{min}, Pa.s	τ_{max} Pa	τ_{min}, Pa	$1/\eta_{max}$, Pa^{-1}.s	$\tau_{max} - \tau_{min}$	CCF***, wt. %
PMMA-Ch*-ZnO	0.18	1.0	1.0	8.5	1.5	1.0	-	
	0.9	1.6	1.2	4.5	1.5	0.63	3.0	
	1.8	1.8	1.4	6.2	1.5	0.55	4.7	5.0
	2.7	58	1.6	7.5	0.75	0.017	6.7	
	3.7	180	10.0	8.6	0.65	0.005	8.0	
	4.3	950	14.0	24.5	0.65	0.001	18.5	
PMMA-BA*-ZnO	0.9	2.0	2.0	28.0	0.5	0.5	-	
	1.8	2.4	2.4	30.0	0.5	0.4	-	
	3.5	85	3.8	35.0	1.5	0.012	33.5	9.0
	5.1	115	4.4	42.0	9.8	0.009	34.5	
	7.0	2800	18.0	55.0	12.8	3.6×10^{-4}	50.0	
	8.0	5500	22.0	67.0	14.0	1.8×10^{-4}	53.0 0	
PMMA-CH*-ZnO	1.8	0.8	0.8	4.5	0.5	1.25	-	
	2.7	0.85	0.85	4.8	0.5	1.18	-	
	3.5	1.5	1.0	4.5	0.8	0.7	3.7	10.5
	5.1	3.8	1.3	5.2	1.2	0.3	4.0	
	7.0	220	2.4	6.8	1.2	0.45×10^{-2}	5.6	
	10.0	520	3.8	18.0	2.5	1.9×10^{-3}	15.5	
PMMA-EA*-ZnO	0.9	2.5	2.3	3.4	0.57	0.4	-	
	1.8	2.9	2.9	3.4	0.57	0.3	-	3.5
	3.5	3.1	3.1	3.4	0.57	0.3	-	

Notes: * Ch – chloroform; BA – benzyl alcohol; CH - cyclohexanon; EA – ethylacetate

** Viscosities η_{max} and η_{min} measured for maximal and minimal possible shear stresses τ and $\dot{\gamma}$ of each dispersion.

*** CCF is filler critical concentration.

Rheology data for deformation of 8 % and frequency of 1 Hz are given in Table 4. They serve for the determination of the viscoelastic properties of sols and gels. Data show that talc provokes decrease of the hardness of gel-like systems. This is a result of the reduction of the density of chain network due to the adsorption of polymer parts on the talc. Besides, the comparison of data for one day and for one month shows that elasticity increases during ageing [63]. Hence, the network evolves in a new structure, forming a larger number of bonds. This effect is weaker for suspensions. We may conclude that talc prevents the polymer chains from structure formation, since adsorption of some segments takes place. Thus, mobility of the adsorbed PAG chains decreases.

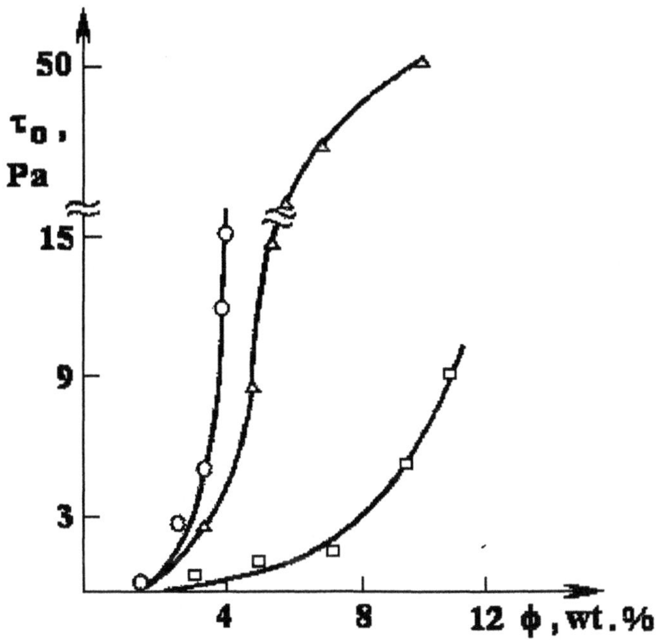

Figure 5. Relations $\tau_0 = f(\phi)$: o - chloroform; Δ - benzyl alcohol; - cyclohexanone

The comparison of test 1 and test 4 shows that the network strengthens with the formation in time of denser junction zones.

[64] illustrates the effect of the interfacial interactions on the rheological behaviour of filled polymer systems. The study considers filled systems consisting on basic in character xanthan and partially hydrolized poly(acrylamide), and acid montmorillonite. The specific features of the process are as follows:

Table 4. Influence of filler content and ageing on the viscoelastic properties of PAG-talc composites

Experiment	Filler content, wt. %	G', N/m^2	G'', N/m^2	G^*, N/m^2	*tan* δ
1. Sol	0	0.92	0.264	0.957	
2. Sol	10	0.594	0.233	0.638	0.287
3. Gel	0	1.35	0.275	1.378	
4. Gel	10	0.715	0.240	0.754	

- as already outlined, in the course of oscillatory measurements the modulus G' attains a plateau value G'_o at low frequencies. The existence of the plateau and the fact that the modulus G'_o is found to be equal to modulus G' proves solid-like behaviour below τ_c.
- the flow curve of xanthan-montmorillonite systems is monotonic; it has a minimum for the partially hydrolized poly(acrylamide) system. If the system has a flow curve with a minimum, it means that we have structure fracture, and its energy of structure fracture is higher than that needed for flow maintenance. The existence of the minimum depends on the viscosity of the suspending medium. Both quantities depend on the polymer content and on the different mechanisms of polymer adsorption on clay particles. Note also that particle aggregation affects the curve profile. Adsorption of xanthan would increase the charge of the clay particle surface, providing a more dispersed structure. For partially hydrolized poly(acrylamide), the flexible molecules would create bridging flocculation [65].

[66] shows how the filler treatment with surfactants affects dispersion rheological properties. The authors use calcium carbonate as a filler with a surfactant-modified surface. The dispersing medium is epoxy resin. The authors indicate that if one performs an optimization of the *filler continuous phase affinity*, the viscosity and pseudo-plasticity decrease. It is clear that an interaction of Lewis acid-base type is advantageous. If chalk is modified with a cationic surfactant, its structure becomes a combination of polar particles covered by a double surfactant layer and non-polar particles covered by a single surfactant layer. The hydrophobic effect initiates external layer adsorption for the two types of particles. The authors assume that the addition of a surfactant to the continuous phase that contains non-treated chalk may produce the same results.

The manufacture of unsaturated polyester resins, used in reinforced and other filled composites, has recently intensified. Most of the unsaturated polyesters, used as binders, undergo contact forming. The popularity of polyesters processed by using highly productive mechanical methods (moulding, contact moulding and extrusion) grows. Note that the car industry, electrical engineering and instrument making are industrial branches where the application of polyesters has the brightest prospects.

A number of recent works [3, 18, 20-22, 28, 38, 67-71] deal with polyester filled composites. The studies try to illuminate the mechanism of gelation of unsaturated polyesters in the presence of metal oxides and hydroxides; they also stress a number of factors that affect gelation rate and depth. Actual problems in the field are the manufacture of non-shrinking polyesters, improvement of the rheological properties of highly filled composites, choice of initiating systems that accelerate composite hardening, etc. There are similar studies dealing with the improvement and control of the material quality and physico-mechanical properties. We will assess the rheological properties of some highly filled polyester composites regarding interfacial interactions.

Viscosity of a polyester composite may increase with the introduction of various fillers – kaolin, $BaSO_4$, $CaCO_3$, etc. One may also introduce some fillers imparting thixotropic properties: aerosil, a finely dispersed silica with strongly developed surface (up to 300 m^2/g - [72]). Bentonite and a number of other compounds cause thixotropy to polyester composites [3]. For extremely high filling, however, the introduction of fillers does not produce highly viscous polyesters with a non-sticky surface. Since polyester viscosity significantly increases, one should not use the compound in the production of prepreg. Moreover, one cannot achieve impregnation of the reinforcement. The thixotropic admixtures do not allow the attainment of the needed rheological parameters during polyester fabrication. All these problems reinforce the theoretical and practical interest in admixtures that chemically react with the polyesters, modifying interfacial interactions and their rheological properties.

As established almost 50 years ago, oxides of a number of two-valence metals from group IIA interact with polyester molecules, and increase the viscosity of polyester systems several times [73]. The importance of this process for technology is that viscosity weakly changes right after the filler introduction, and then significantly increases in the process of material preparation, attaining large values, as shown in Fig. 6.

The figure shows that η follows an *S*-shaped curve. There are three stages of viscosity modification: stage 1 – viscosity does not change significantly after the introduction of the metal oxide; stage 2 – sharp increase of viscosity; stage 3 – viscosity becomes constant (of magnitude 104 – 106 Pa.s). This character of η is extremely important for the production of a prepreg, since it provides uniform and fast impregnation of the reinforcement.

[73] employs model systems in their early studies on polyesters. They consider real polyester systems only in his later works. Firstly they choose ZnO as a solid base. It can be noted that one may achieve significant modification of composite rheological properties, even when treating polyesters that do not contain acid end carboxyl groups. However, it is necessary to introduce an equivalent quantity of acetic acid together with ZnO. They explain the phenomenon by the formation of complexes with the carboxyl groups.

It is important to note that by using MgO as a filler, the composition viscosity and the concentration of the disssolved magnesia are related as follows [74]:

$$\eta = \eta_0 \, e^{\alpha c}, \tag{3.16}$$

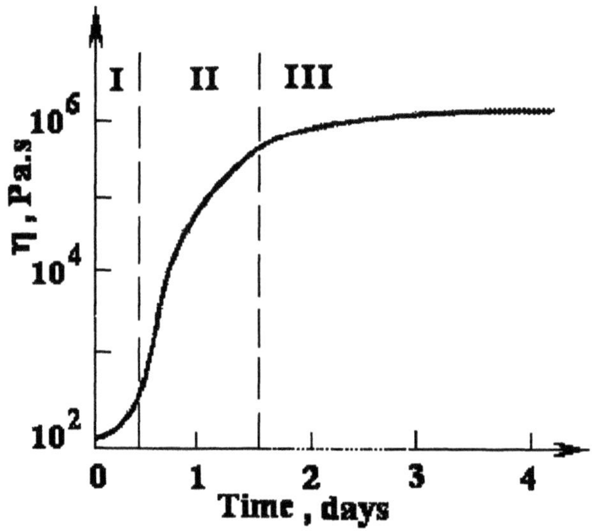

Figure 6. Kinetics of the increase of viscosity of a polyester resin filled with metal oxides

where η_0 and α are constants and their values depend on the property of unsaturated polyester and on the concentration of the dissolved MgO. The study [74] shows that small change of the MgO concentration yields significant change of the viscosity – Fig. 7.

Fig. 7 also shows that the concentration of the dissolved magnesia is larger than that for the MgO, needed for the formation of a neutral salt (0.41 % for a resin with acidic number 19). Hence, some quantity of a basic salt forms. The author assumes that the formation of Mg salts during the first stage takes place relatively rapidly, and does not change the viscosity significantly. The second stage covers the formation of coordination bonds between the metal salts and the carbonyl groups of the polyester; this stage is slower and more extensive than the first, and is characterised by a significant increase of viscosity.

[74] sets forth another two-stage mechanism of viscosity modification, proposed by Alvey. The study considers polyester resins filled with metal oxides. It outlines the effect of the ratio *polyester carboxyl groups/metal oxides (hydroxides)* on polyester viscosity. Using magnesium oxide and

sodium hydroxide, the author finds that NaOH does not affect viscosity of the polyester resin, even for 100% neutralization of the polyester acidic groups (Fig. 8).

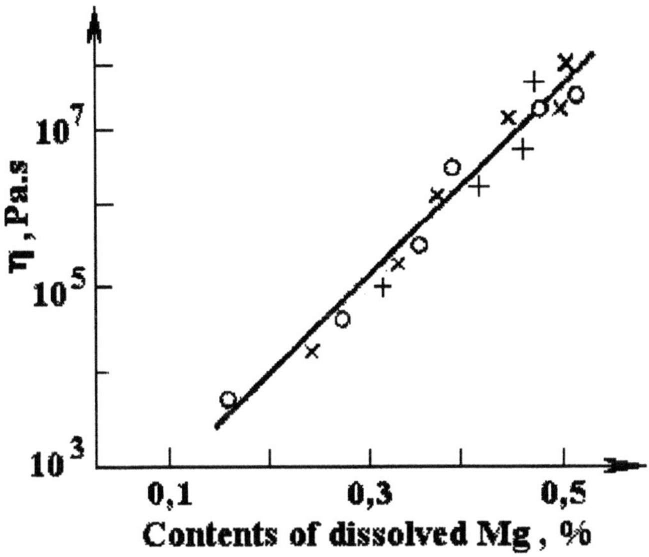

Figure 7. Dependence of the viscosity of a polyester resin on the content of dissolved magnesia, wt. %: o – 1, x – 1.5, + – 2

The author also notes that the excess of sodium hydroxide causes an increase in the viscosity comparable to that arising from the introduction of magnesium oxide. The ion interaction between the polyesters and the metal bases gives proof of this phenomenon. The results show that viscosity change depends on the equivalent ratio MgO/COOH. Besides, viscosity attains a maximum for a ratio 3:1, and further increase of the MgO concentration does not change the limit values. Hence, one gets a relation similar to that of stages II and III, indicated in Fig. 6. The author supposes that an equilibrium reaction of metal oxides or hydroxides with the polyester carboxyl groups takes place during the first stage, leading to salt formation. The second stage reflects interchain dipole interaction of the polymer salts with the excess of metal compounds, and yields a significant change in the viscosity.

We note that a one-stage mechanism of viscosity modification is also popular; it excludes the stage of complex formation [75]. The authors suppose that three types of polyester molecules form during the process: molecules with a single hydroxyl and a single carboxyl group (OH ~~ COOH), molecules with two hydroxyl groups (OH~~OH), and molecules with two carboxyl groups

(HOOC~~COOH). If molecules of the first type participate in the salt formation, the molecular weight may only double and this should not yield an essential increase of η. If however, molecules of the third type participate in the reaction, the molecular weight may attain significant value due to polycondensation with the metal oxides and salt formation. Those polyester structures cause significant change of the system rheological properties. Based on such assumption, the authors model the kinetics of polyester viscosity modification. They consider a polyester resin filled with oxides of alkaline earth metals. The process comprises solid-liquid diffusion of the metal oxide, reaction of the oxide with the polyester carboxyl groups yielding a basic salt, and further interaction of this basic salt during the increase of η. The authors show good agreement between the calculated and experimental rate of viscosity increase. Besides, the relation $ln(\eta/\eta_0) = f(t)$ is similar to that of stages II and III shown in Fig. 6.

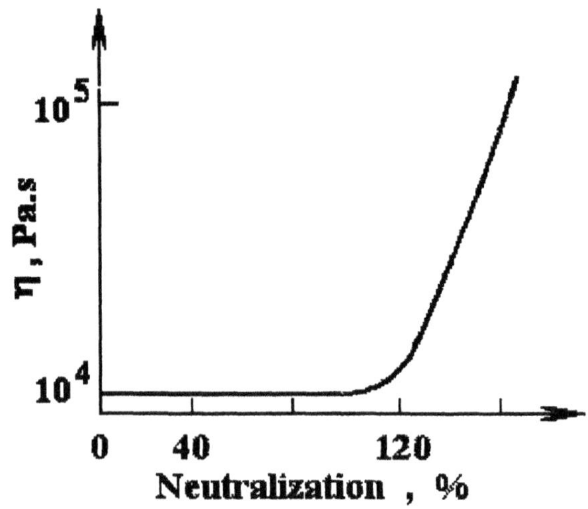

Figure 8. Change of η of a polyester resin during neutralization of its carboxyle groups by introducing bases

The authors of [73-75] note the essential part of water in the process of polyester thickening (i.e. in the increase of η). They link the positive effect of water (a compound with a high dipole moment) with the breaking of the hydrogen bonds that form on polyester carboxyl groups.They find that water increases the initial rate of viscosity change and considerably decreases the final limiting value. Note that some authors explain this phenomenon by the improvement of MgO solubility within the reaction medium; they link the decrease of the limiting value of η with the hydrolysis of magnesium salts, observing that the limiting value most strongly decreases for water contents up to 1 %.

It is worth noting that in contrast to systems with MgO, water in systems with ZnO or $Ca(OH)_2$ accelerates polymerization and increases viscosity – see Table 5. This holds true even for high water content – up to 5 %.

Table 5. Dependence of viscosity of a polyester resin filled with ZnO vs. the water content, $T = 45$ °C

H_2O content, %	η in cPx10^{-4} at t, hours				
	1	2	3	4	5
0.75	1.0	2.0	2.5	3.0	25
1.0	1.5	4.0	6.0	12.15	-
2.0	2.5	6.2	16.25	17.5	-
5.0	4.0	9.5	-	-	-

The analysis of these studies shows that besides the formation of complexes including carboxyl groups, there may also be formations of aqua-complexes in the polymer systems. The increase of the water content, which is present in a system with MgO, shifts the system equilibrium in favor of the formation of aqueous complexes. Complex formation takes place on the carboxyl groups, yielding structuring of the polyester resins; this suppresses the system and prevents viscosity increase. The zinc oxide is inclined to form complexes on the carboxyl groups, and in this case the increased water content stimulates ZnO solubility in the system.

[73] discusses various oxides and hydroxides of alkali earth metals and some salts as chemically active systems. Those are the well known solid bases MgO, $Mg(OH)_2$, CaO, $Ca(OH)_2$, ZnO, BeO, PbO, Li_2O, LiOH, BaO, $Ba(OH)_2$, K_2O, Na_2O, NaOH, salts of Al and Ga, organic acids, etc. The existence of such a wide set of chemically active admixtures - thickeners, together with their combinations, reveal significant possibilities for viscosity variation during the composite (prepreg) preparation. The authors propose their ordering, based on the rate of viscosity increase: MgO $\approx$ $Ca(OH)_2$ $>$ CaO $>$ $Mg(OH)_2$. Based on the maximal viscosity value, this order is as follows: MgO $>$ $Ca(OH)_2$ $>$ $Mg(OH)_2$.

We have already noted that one may obtain a definite effect by using a combination of thickeners. Literature provides a number of possible combinations. We give here two of them – CaO/$Ca(OH)_2$ or CaO/MgO [73] in Table 6.

One may use highly effective systems to diminish the thickening rate at stage 1 and to accelerate this rate during the following stages; the systems are usually based on MgO (being basic in character) and lithium salts (acidic in character) [73], Table 7.

Table 6. Influence of thickeners on the viscosity of a polyester resin

Content, %			η, cP.10^3 at time, hours				
CaO	**Ca(OH)₂**	**MgO**	**1**	**3**	**5**	**24**	**168**
-	5.0	-	26.5	75.0	116.0	380	3000
3.0	2.0	-	1.5	1.6	2.1	650	8600
-	-	3.6	6.5	79.0	180.0	1940	10200
2.5	-	1.8	3.3	14.7	37.0	1100	42000
3.8	2.9	-	1.65	2.0	76.0	5800	25000

Table 7. Effect of the filler quantity on viscosity of polyester resins

Quantity, wt. parts		η, cP.10^3 at *t*, hours							
Mg O	**LiCl**	**4**	**7**	**24**	**30**	**48**	**72**	**96**	**192**
1.0	0	5.0	6.1	22.5	30.5	70.0	146	272	832
1.0	0.1	5.0	5.6	18.0	25.0	79.0	240	288	4500
1.0	0.7	5.0	5.5	23.0	44.0	316.0	1440	5120	22500
1.0	1.0	4.9	5.3	25.0	48.0	268	1296	5312	19200
1.5	0	6.2	9.0	67.0	120	380	928	1840	6528
1.5	0.1	6.1	7.8	45.7	81.0	256	896	3163	13800
1.5	0.7	5.8	7.0	40.0	80.5	810	4160	13760	46000
1.5	1.0	5.8	6.6	49.5	124	1220	8832	19200	58000
2.0	0	5.5	7.4	68.0	130	480	1184	2368	9600
2.0	0.1	5.5	7.0	39.5	66.5	280	1984	6276	23600
2.0	0.7	5.4	6.5	51.0	112	1540	10240	26880	64000
2.0	1.0	5.0	6.5	70.0	292	3390	16640	31360	64000

[73] presents a generalization of a number of patents dealing with the amphoteric character of polyester resin. The authors discuss a number of compounds of acidic type as effective thickening accelerators – HCl, HBr, $POCl_3$, $SOCl_2$, $SiCl_4$, $SnCl_4$, $TiCl_4$, $AlCl_4$, HB_3. Besides, they consider some organic acids – metacrylic, benzoic, acetic, chlorous-acetic, acrylic, phenyl-acetic, phenoxiacetic, cyclohexane-carbonic, etc., as well as anhydrides of cyclic carbonic acids - phthalic, tetra- and hexahydrophthalic, etc. One may find more details on various accelerators and regulators of thickening (viscosity increase) in patent literature and in [73].

As we have shown, the compounds require regular thickening (viscosity change) in time. This is a requirement for compositions used for the preparation of prepreg and premixes. Engineering ground works and the manufacture of pipes, industrial pans, etc. require the use of polyester preventative coatings or gums. One may also use those compounds as concrete binders. Note that technology usually combines polyester preparation with the fabrication of polyester products. The main technological requirement, however, is to minimize material and energy. Besides, the properties of the technological mixes and the exploitation characteristics of the final product should also suit definite requirements. At the same time, the variety of technologies and the complexity of the structure formation of filled polymers do not allow for an adequate quantitative description of the compounds. Voznesenski, Ivanov and co-workers [28, 76, 77] propose relations for the control of composite properties; they employ experimental-statistical models. Their studies provide an effective analysis of the filler effect on polymer structure formation and rheological properties of compositions. The authors use a model of the type *compound I – compound II – technology - properties* ($C_I - C_{II} - T - P$). The plan (design) of the subsequent experiment involves the following components: *mix I* – fillers that have different acid-base surface properties – V_1, V_2 and V_3; *mix II* – fillers of various dispersity as determined using Blain's method – W_1, W_2 and W_3; *technology* – a characteristic of external impacts; *properties* – in this case they are the viscosities of liquid composites.

Materials used for mix I are V_1 – solid acid SiO_2, V_2 – solid base CaO and V_3 – amphoteric complex filler containing SiO_2 – 20.4 %, Al_2O_3 – 5.1 %, CaO – 64.6 %, Fe_2O_3 – 3 %, MgO – 1.5 %. We choose those popular fillers from the set of admissible ones [78 – 82]. Experimenting with different types of fillers, we aim at obtaining a significant variability of the composite properties.

Materials for mix II are the same as those for mix I, but have different dispersity: W_1 – finely dispersed particles with a specific surface $S_1 = 332 \pm 19$ m^2/kg; W_2 – particles with medium size where $S_2 = 97 \pm 7$ m^2/kg and W_3 – particles with large size where $S_3 = 46 \pm 4$ m^2/kg. We analyze nine powder fillers, following the outlined filling scheme. Here:

- V_1 – SiO_2 with S_1, S_2 and S_3;
- V_2 – CaO with S_1, S_2 and S_3;
- V_3 – complex filler with S_1, S_2 and S_3 .

Choosing the appropriate filler, we obtain not only three fillers of different acid and/or base activity, but also nine fillers with different specific surface, able of interacting with the polymer. The polymer used is unsaturated polyester resin – a product of the polycondensation of equal parts of phthalic and maleic anhydride with ethylene and propyleneglycol. The compound, which molecular weight is 2800 and density – 1.12 g/cm^3, is dissolved in styrene. To design the model, we use an experimental plan that enables one to assess 45 parameters of the model $C_I - C_{II} - T - P$. The plan of the experiment comprises 45 experimental points which correspond to 28 different filler compositions – Table 8.

We choose the control points 101 ... 109, taking account of the symmetry condition. Based on the experimental results, we design models that describe the change of $S = f(V_i, W_i)$ and $\eta = f(\dot{\gamma})$ [28].

The experiments for filler/polymer ratio of $F/P = 1.5$ and $F/P = 2.5$ show that the prepared liquid composites display pseudoplastic behaviour. A hyperbolic function, corresponding to Oswald-de-Willa's model, maps the relation $\eta = f(\dot{\gamma})$ [3, 83]. For high values of $\dot{\gamma}$ the structure is destroyed, and the composite flow becomes Newtonian, while η gets a minimal value η_m. The characterization of the filler type and filler dispersity is carried out by designing the following model (note that quantity $\dot{\gamma}$ is fixed, i.e. $\dot{\gamma} = const$ [3, 28, 83]):

Table 8. Plan of the experiment for C_I - C_{II} – T - P model

No. of filler	**Factor levels**						**Experiment number for** x		
		Mix I			**Mix II**		**-1**	**0**	**+1**
1	1	0	0	1	0	0	2	101*	32
2	1	0	0	0	1	0	3	19	103*
3	1	0	0	0	0	1	4	17	34
4	1	0	0	1/2	1/2	0	102*	20	33
5	0	1	0	1	0	0	104*	22	37
6	0	1	0	0	1	0	7	23	35
7	0	1	0	0	0	1	5	21	105*
8	0	1	0	1/2	0	1/2	6	106*	38
9	0	0	1	1	0	0	8	26	41
10	0	0	1	0	1	0	107*	24	39
11	0	0	1	0	0	1	10	109*	40
12	0	0	1	0	1/2	1/2	11	25	108*
13	1/3	1/3	1/3	1/3	1/3	1/3	1	16	31
14	1	0	0	1/2	0	1/2	-	18	-
15	0	1	0	0	1/2	1/2	-	-	36
16	0	0	1	1/2	1/2	0	9	-	-
17	1/2	1/2	0	1	0	0	14	-	-
18	1/2	1/2	0	1/2	0	1/2	-	29	-
19	1/2	1/2	0	0	1/2	1/2	-	-	44
20	1/2	0	1/2	0	1	0	-	-	43
21	1/2	0	1/2	1/2	1/2	0	13	-	-
22	1/2	0	1/2	1/2	0	1/2	-	28	-
23	0	1/2	1/2	0	0	1	-	27	-
24	0	1/2	1/2	1/2	1/2	0	12	-	-
25	0	1/2	1/2	0	1/2	1/2	-	-	42
26	1/3	1/3	1/3	1		0	-	30	-
27	1/3	1/3	1/3	0	1	0	15	-	-
28	1/3	1/3	1/3	0	0	1	-	-	45

Note: * Points 101 ... 109 may serve as control points during dense experiment.

$$\begin{aligned} \ln \eta = {} & (4.51 + 1.24x - 0.13x^2)V_1W_1 + (4.24 + 0.86x + 0.10x^2)V_1W_2 + \\ & + (4.90 + 0.82x - 0.35x^2)V_1W_3 + (5.46 + 1.23x - 0.66x^2)V_2W_1 + (5.01 + \\ & 0.83x - 0.44x^2)V_2W_3 + (4.83 + 0.71x \pm 0.24x^2)V_2W_3 + (4.52 + 1.49x \pm \\ & \pm 0)V_3W_1 + (0.53 + 1.11x \pm 0)V_3W_2 + (4.27 + 1.04x + 0.28x^2)V_3W_3 + \\ & + (-0.64 + 0.70x + 0.54x^2)V_1V_2 + (\pm 0 - 1.14x + 0.56x^2)V_1V_3 + \\ & + (-0.94 - 0.33x \pm 0)V_2V_3 + (-1.43 - 0.63x \pm 0)W_1W_2 + (-2.58 - \\ & -0.42x \pm 0)W_1W_3 + (-1.04 \pm 0 \pm 0)W_2W_3 \end{aligned} \quad (3.17)$$

Model (3.17) describes the system mix I (acid-base surface properties) – mix II (dispersity) – technology (filling degree) – properties (composite viscosity for $\dot{\gamma}$ = 1 s^{-1}). The parameters have definite physical meaning, providing the overall physics of the system [28]. Thus, the function in the brackets after the first term V_1W_1 gives the parabolic change of η - a composition filled with finely disperse SiO_2 particles. Here the degree of filling varies. The free term corresponds to the viscosity of the composition for F/P = 2 (x = 0). The coefficients in front of x and x^2 give estimates of the rate and acceleration of viscosity decrease with the decrease of the filling degree. Consider the viscosity logarithm; the polynomials of x in front of the terms V_iV_j and W_iW_j express the non-linearity, and non-additivity of mixing fillers and particles of different specific surface. The polynomials reflect also the viscosity value dependence on the filling degree. [28, 83] present similar models for all possible and considered levels of change of $\dot{\gamma}$.

The proposed model (3.17) enables one to build a set of *composition-properties* diagrams of the type *acid-base surface properties-viscosity* and *dispersity-viscosity*.

So far we have analyzed the use of CaO as a component of polyester composites used for the preparation of prepreg and premixes. Now we consider how the actual dispersity (or a combination of dispersity) of CaO affects viscosity. We base our considerations on the model (3.17). When putting V_2 = 1, $V_1 = V_3 = 0$ and x =1 in the model (x_1), we get the so-called model pattern:

$$\ln \eta_1 = 6.03W_1 - 5.40W_2 + 5.29W_3 - 2.06W_1W_2 - 3.0W_1W_3 - 1.04W_2W_3 \quad (3.18)$$

Then, the isolines of η of the CaO composition with F/P = 2.5 have the shape shown on the dispersity-specific surface triangle (W_1, W_2, W_3) (Fig. 9); the characteristic change of η, being affected by the dispersity. The singularities $M_1 ... M_6$ are clearly seen. Table 9 presents the corresponding values of η_1, W_i, S and δ_p - the calculated average thickness of the polymer layer that covers the solid surface:

$$\delta_p = 10[S(F.P)\rho_p]^{-1}, \qquad (3.19)$$

where ρ_p is the polymer density.

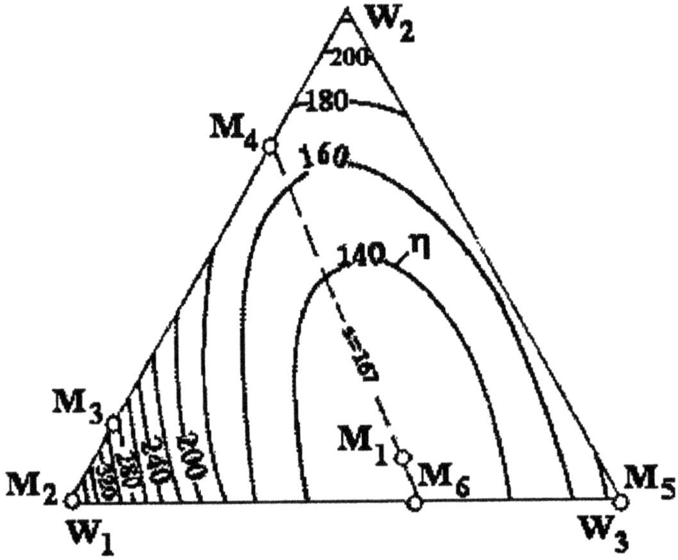

Figure 9. Diagram "*dispersity- η*" and characteristic points given in Table 9 for systems with CaO fiiller, $\phi = 2.5$ wt. % and $\dot{\gamma} = 1\ s^{-1}$

Table 9. Characteristic points on the dispersity-viscosity diagram for the CaO filled polyester composite ($\dot{\gamma}$ = 1 s-1, *F/P* = 2.5)

Points on Fig. 9	Filler			*S*, m^2/kg	δ_p, μm	*η*, Pa.s
	W_1	W_2	W_3			
M_1	0.35	0.09	0.54	167	2.1	128
M_2	1	0	0	335	1.0	412
M_3	0.85	0.15	0	300	1.2	291
M_4	0.72	0.28	0	167	2.1	174
M_5	0	0	1	48	7.3	198
M_6	0.39	0	0.61	171	2	129

We generalize the viscosity values at the singularities, together with other numerical characteristics of the dispersity effect G_ω. The following parameters characterize viscosity η as a function of the filler dispersity:

- **change** of viscosity from a maximal value (for a finely dispersed filler) to a minimal value (for fraction mixes):

$$\chi_{max} = \eta_1\{M_2\}/\eta_1\{M_1\} = exp\,[ln\,\eta_1\{M_2\} - ln\,\eta_1\{M_1\}] \tag{3.20}$$

- index caracterizing the **decrease** of η during a transfer from a filler that yields maximal viscosity (η_1^{300}) for $S = 300$ m^2/kg, to a filler that provides minimal viscosity for a subsequent specific surface:

$$\chi^{300} = \eta_1\{M_3\}/\eta_1\{M_1\} = exp\,[ln\,\eta_1\{M_3\} - ln\,\eta_1\{M_1\}] \tag{3.21}$$

- **averaged gradient**, characterizing the rate of viscosity decrease from a maximal value $\eta_{1\,max}$ to a minimal value $\eta_{1\,min}$, under a decrease of the filler surface S to 100 m^2/kg

$$\nabla_{max} = (\chi_{max} - 1)/[S\{M_2\} - S\{M_1\}] \tag{3.22}$$

So far we have studied the effect of CaO dispersity. We shall now comparatively assess the effect of filler surface activity, i.e. that of filler acid-base surface properties ($V_1 + V_2 + V_3$). The model obtained from (3.17) for $x = +1$ describes the set of filler acid-base surface properties, i.e. the surface activity. The value $x = +1$ corresponds to the upper filling level within the studied range $F/P = 2.5$. The viscosity relation reads:

$$ln\,\eta_1 = 5.62V_1W_1 + 5.20V_1W_2 + 5.37V_1W_3 + 6.03V_2W_1 +$$

$$+ 5.40V_2W_2 + 5.30V_2W_3 + 6.01V_3W_1 + 5.46V_3W_2 +$$

$$+ 5.59V_3W_3 + 0.60V_1V_2 - 0.58V_1V_3 - 1.27V_2V_3 - 2.06W_1W_2 -$$

$$- 3.00\,W_1W_3 - 1.04W_2W_3 \tag{3.23}$$

To map the varying three-component *dispersity-viscosity* diagrams, we use a discrete set of seven points – centroids of the *acid-base surface properties-viscosity* basic triangle (Fig. 10). The *dispersity-viscosity* diagram for CaO occupies the V_2 peak of the carrying basic triangle. The generalized parameters produce the character of the isolines for η. The triangle diagram ($V_1 - V_2 - V_3$) shows that a range of minimal viscosity exists for a triple mixture of fillers. Within this range one may obtain minimal viscosity by combining the three fillers, but each of them should have one and the same dispersity. We calculate them for all seven diagrams shown in Fig. 10. Thus, we get the values for S (m^2/kg) corresponding to the minimal values of η at points M_1 for :

SiO_2 ($V_1 = 1$) - 182

Binary mixture $(0.5V_1 + 0.5V_2)$	- 182	
CaO $(V_2 = 1)$	- 167	
Binary mixture $(0.5V_2 + 0.5V_3)$	- 142	(3.24)
Polymineral filler $(V_3 = 1)$	- 154	
Binary mixture $(0.5\ V_1 + 0.5\ V_3)$	- 168	
Triple mixture $(V_1 = V_2 = V_3 = 1/3)$	- 163	

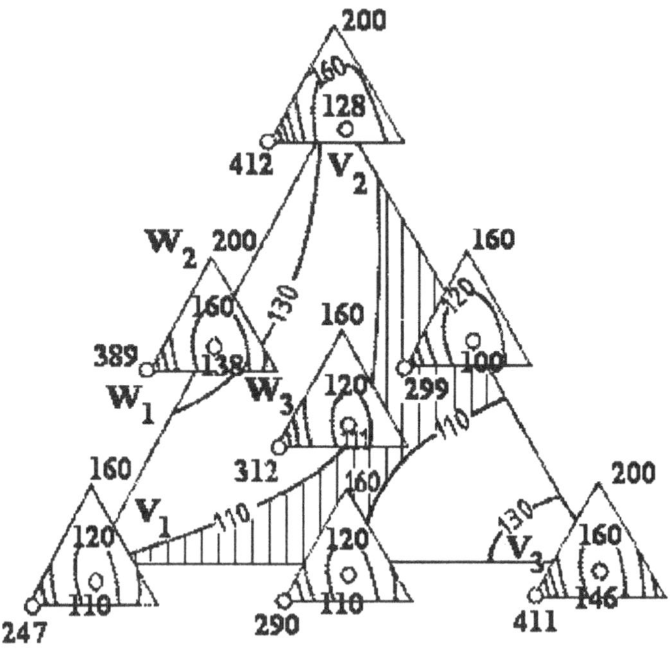

Figure 10. "Dispersity - viscosity" diagrams and isolines corresponding to the minimal values η_{min} of the acid-base surface properties triangle

The results of (3.24) correspond to those obtained by using the methods in [28]. One may generalize the results by designing a new model $S\{M_1\} = S\{\eta_{min}\} = f(V_1)$, using Scheffe's polynomial:

$$S\{M_1\} = 182V_1 + 167V_2 + 154V_3 + 30V_1V_2 + 0V_1V_3 - 74V_2V_3 \qquad (3.25)$$

We get $S = 162.9\ m^2/kg$ from the model (eqn. 3.25). The value of S coincides with row 7 of relation (3.24). Fig. 11 shows the isolines of the diagram acid-base surface properties - $S\{\eta_{min}\}$.

We may use viscosity η^{300} as a comparative characteristic of the filler surface activity. We obtain the viscosity value by solving eqn. (3.21). Here viscosity stands for an integral characteristic of the filler structure-formation capability. Note [84, 85] that structure formation is due to adsorption, chemical bonding, dispersity, etc. Under *F/P = const.* = 2.5; *S = const.* = 300 m^2/kg and $\dot{\gamma}$ = *const.* = 1 s^{-1}, the composition with a complex filler is most able to form a structure. Its viscosity is η_1^{300} = 352 Pa.s. The ability of calcium oxide for a structure formation is 20 % lower (η_1^{300} = 291 Pa.s). Composites with SiO_2 have the lowest capability for a structure formation, where η_1^{300} = 181 Pa.s. Those results correspond to the strength of interfacial interactions that take place within a polyester resin with different fillers. Note that the polyester resin has amphoteric properties.

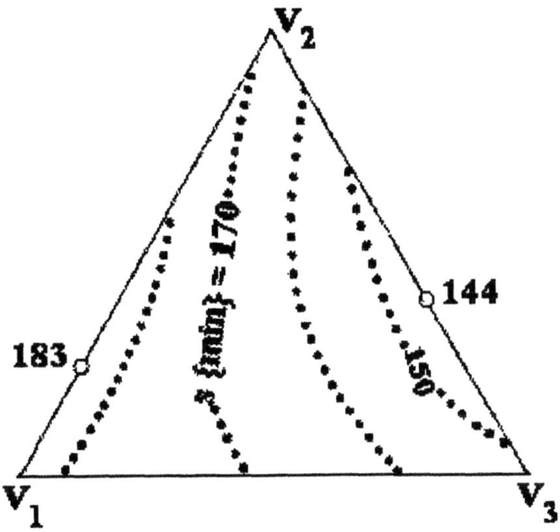

Figure 11. Acid-base surface properties diagram of the generalized parameter $S\{\eta_{min}\}$

We should note, however, that the structure-formation capability is not additive for triple mixes (Fig. 11). This is true especially for compositions that contain complex filler, as shown by the model:

$$\ln \eta_1^{300} = 5.200V_1 + 5.672V_2 + 5.870V_3 + 0.142V_1V_2 - 0.660V_1V_3 - 0.640V_2V_3 \quad (3.26)$$

As we have shown, the viscosity of mixes, containing complex filler (V_3), decreases. This is due to the slower increase of viscosity during the first hours of composite preparation in the presence of MgO [73]. One may use the specific behaviour of viscosity to optimize the filler quantity. Fig. 10 confirms this technological conclusion; it presents the carrying acid-base surface properties diagram with isolines of minimal viscosity $\eta_{1\,min}$, obtained during the control of composite dispersity.

The parameter χ^{300} - eqn. (3.21), is the viscosity decrease from η_1^{300} to $\eta_{1\,min}$. We use it here to characterize the stability of structures formed by adding different mineral powders. Note that this parameter depends only on the filler acid-base surface properties for *F/P = const.* (Fig. 12), i.e. it depends on a filler activity during the interaction with the polymer. In this case the stability of the structure is considered with regard to the increase of the thickness of the polymer layer as the specific surface of the filler decreses. The phase interactions are most active within a range η_1^{300} and $\eta_{1\,min}$, i.e. point M_3 - point M_1. Hence, the structure is stable and the increase of the thickness of the polymer layer is natural. Regarding the hydrodynamic theory of lubricants [28], decrease of viscosity accompanies the thickness increase. One may consider the specific surface $S\{M_1\}$ as a conventional border of the range of stable structures. Note that the specific surface corresponds to the minimal viscosity $\eta_{1\,min}$ - Fig. 9.

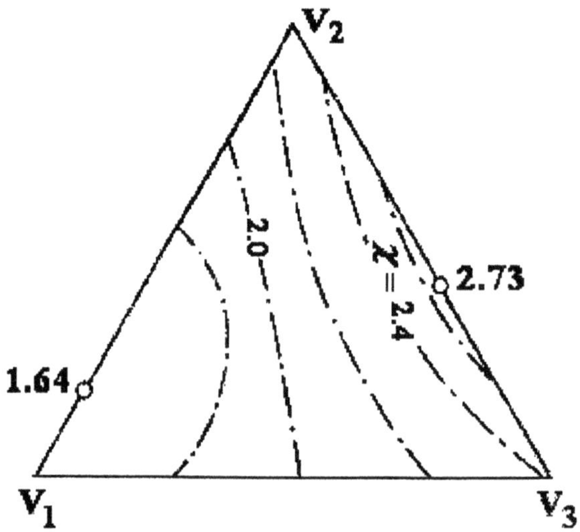

Figure 12. Acid-base properties diagram of the generalized parameter η_1^{300}

Fig. 12 shows that composites containing large quantity of SiO_2 ($V_1 \geq$ 0.6…0.7) have minimal stability. Besides, the dispersity $S\{M_1\}$ should increase to keep composite stability – see Fig. 12. The *acid-base surface properties-structure stability* χ^{300} diagrams and *acid-base surface properties-S{M₁}* diagrams are close to each other. Their character corresponds to the relation between the various structure parameters. The relation is close to inverse proportionality. Considering the value χ^{300}, structures with CaO and with complex filler have approximately the same stability. Their stability is much higher than that of a structure with SiO_2. Structures with CaO and complex filler (i.e. $V_2 + V_3$) gain maximal stability for $S\{M_1\} = 141$ m^2/kg. This value is almost half the value for SiO_2.

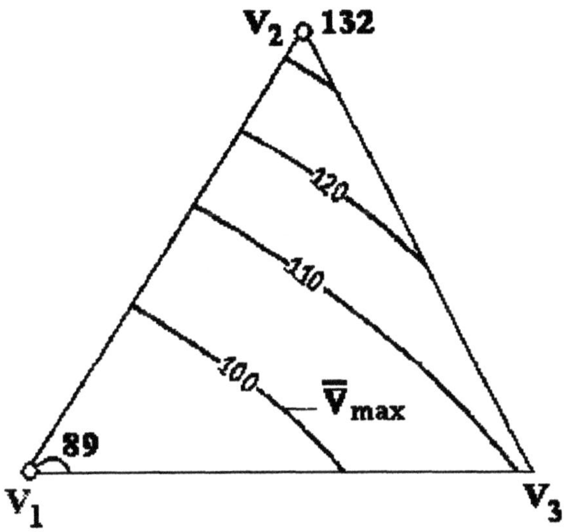

Figure 13. Acid-base properties diagram of the generalized parameter $\overline{\nabla}_{max}$

The similarity between the diagrams in Fig. 11 and Fig. 12 and the results of the analysis prove that we should proceed with the generalization when tackling the structure-formation part of the filler. We may use the gradient ∇_{max} (eqn. 3.22) as a specific characteristic. As designed, the gradient comprises structure-formation capability and other parameters: minimal viscosity $\eta_{1\ min}$ attained by dispersity control, structure stability χ^{300} and the limit dispersity $S\{M_1\}$. The gradient is shown in Fig. 13, which shows that the minimal gradient characterizes compositions with $Si0_2$, while the maximal one – compositions with pure CaO. Hence, we may suppose that CaO is a compound that strongly stimulates the structure formation in polymer liquid layers. However, the degree of structure formation takes the largest value for $F/P = const. = 2.5$ and $\dot{\gamma} = const. = 1$ s^{-1}. Note that those considerations are valid for a wider set of compounds, regarding $S = 100$ m^2/kg and $S = 200$ m^2/kg of different mineral fillers and filler mixes – see Fig. 14 and Fig. 15.

The analysis of the *acid-base surface properties-gradient isoparametric decrease of the viscosity* (∇^{100} *or* ∇^{200}) diagrams proves that polyester composites with CaO are most sensitive to the change of the distribution of particle sizes. The complex composite is less sensitive to that characteristic, which seems reasonable, since the overall contents of CaO and MgO are small. Least sensitive to the particle distribution is polymer with SiO_2. Thus, the additional parameter of the filler structure formation capability ∇^s shows that minerals of higher surface activity significantly increase the composite sensitivity to optimal arrangement of the filler particles.

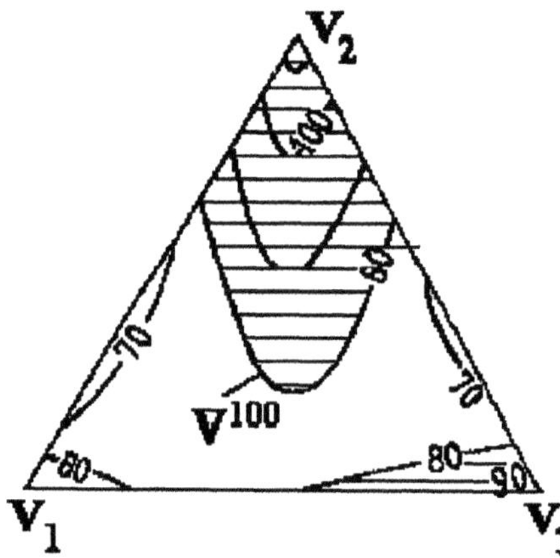

Figure 14. Change of η_l for $S = 100$ m²/kg, $F/P = 2.5$ regarding mixtures of SiO_2 (V_1), CaO (V_2) and complex filler (V_3) (∇^{100})

We base our considerations on *dispersity-viscosity* diagrams, produced by the acid-base surface properties triangle – Fig. 10. The same results are valid for the opposite system, i.e. for *acid-base surface properties-viscosity* diagrams that are given by the triangle of dispersity – Fig. 16. It is seen that the character of the *acid-base surface properties-*η_l^∞ diagram is close to the maximal viscosity shown in Fig. 10. That is especially true for mixes of particles of different fractions. Moreover, the isolines of $\eta_{l\ min} = f(V_i)$ of the basic triangle shown in Fig. 16 are similar to those of the seven triangles shown in Fig. 10. They are the closest ones to the diagram of the $V_3 + V_2$ mixes [83].

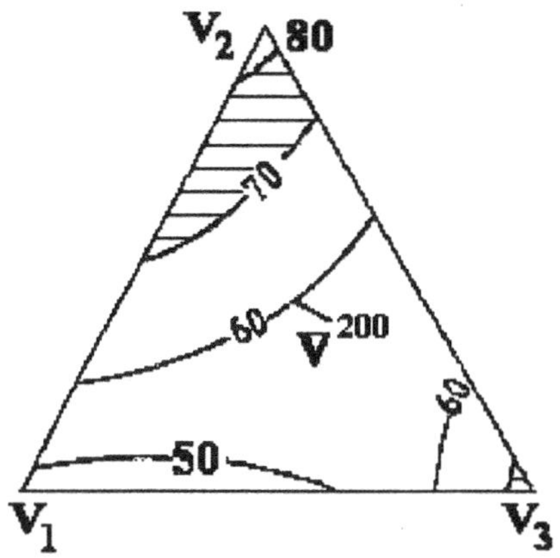

Figure 15. Change of η_1 for $S = 100\ m^2/kg$, $F/P = 2.5$ regarding mixtures of SiO_2 (V_1), CaO (V_2) and complex filler (V_3) (∇^{200})

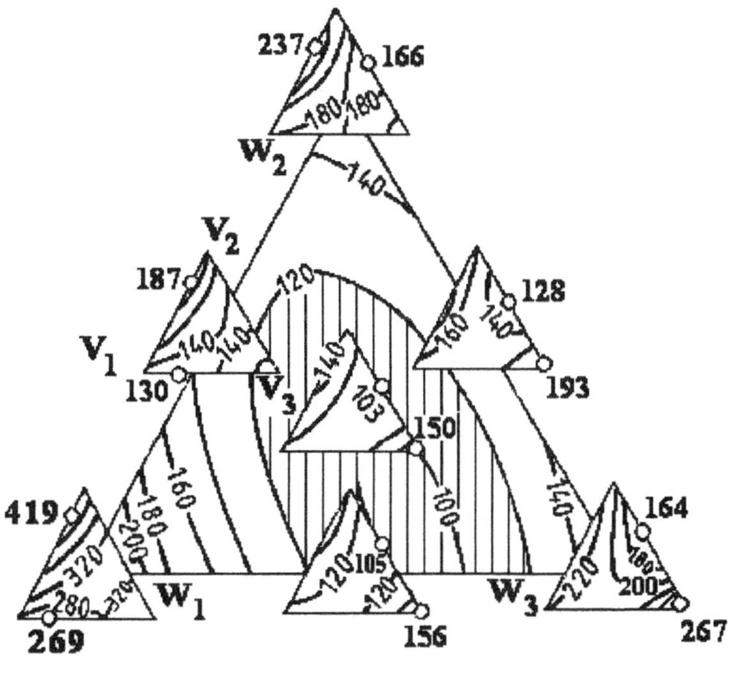

Figure 16. Acid-base surface properties - viscosity diagrams and isolines corresponding to the minimal values η_{min} of the basic triangle of dispersity

The above-cited data and those submitted in [28] show how the filler activity and dispersity can affect composite rheological characteristics at a constant filling degree. They are used for choosing an optimal filler type for the polymer composites.

REFERENCES

1. Reiner, Markus, *Rheology*. Moscow: Mir, 1965 (In Russian).
2. Malkin, Alexander, *Rheology Fundamentals*. Ontario: Chem. Tec. Publ., 1994.
3. Ivanov, Yatchko, *Rheology and Structure of Liquide Silicate and Polymer Composites*. Sofia: BAS, 1988 (In Bulgarian).
4. Lenk, Robert, *Polymer Rheology*. London: Appl. Sci. Publ., 1978.
5. Cross, Malcolm. "A flow of Pigment-Oil Suspensions of the Printing Ink Type." In *Polymer Systems. Deformation and Fflow*, Wetton, H. and Whorlow M., eds. N. Y.: Macmillan, 1968, 124-43.
6. Carreau P. Rheological equation from molecular network theories. Trans. Soc. Rheol. 1972; **16**: 99-127.
7. Yasuda K.Y., Armstrong R.C., Cohen R.E. Shear flow properties of concentrated solutions of linear and star branched polymers. *Reol. Acta* 1981; **20**: 163-78.
8. Shulman, Zinovii, *Convective Heat Transfer in Rheologicaly Complex Fluids*. Moscow: Energia, 1975 (In Russian).
9. Van Oene H. "Rheology of polymer blends and solusions." In *Polymer Blends*, vol. 1, D. R. Paul and S. Newman, eds. Moscow: Mir, 1981, 337-96.
10. Cirman C., Bonquey M., Terrisse J., Muller R. "Melt rheology of reactive polymer blends." In *Progress and Trends in Rheology*, Igor Emri, ed. Darmstadt: Elsevier, 1998, 381-2.
11. Chang C., Powell R.L. Effect of particle size distribution on the concentrated bimodal suspensions. J. Rheology 1994; **38**: 85-8.
12. Laun H.M., Bung R., Hess S., Loose W., Hess O., Hahn K., Hadicke E., Hingman R., Schmidt F., Lindner P. Rheological and small angle neutron scattering investigation of shear-indused particle structure of concentrated polymer dispersions. J. Rheology 1992; **36**: 743-8.
13. Phan-Thien N. Constitative equation for concentrated suspensions in Newtonian liquids. J. Rheology 1995; **39**: 679-95.
14. Probstein R., Sengun M., Tseng T.-C. Bimodal model of concentrated suspension viscosity for distributing particle size. J. Rheology 1994; **38**: 811-29.
15. Ranganthan S., Advani S. Characterization of orientation clustering in short - fiber composites. J. Polym. Sci. Part B. Polym. Physics, 1990; **28**: 2651-72.
16. Shikata T., Pearson D. Viscoelastic behaviour of concentrated suspensions. J. Rheology 1994; **38**: 601-16.
17. Yanovsky Yu. "Rheological and rhelaxation properties of polymer composites with fillers." Proceedings of the International Conference on Mechanics and Technology of Composite Materials; 1991 October 14 - 16; Sofia: BAS, 1991, 46-49 (In Russian).
18. Ivanov Ya. Microrheology of composite materials. Proceedings of the International Conference on Mechanics and Technology of Composite Materials; 1991 October 14 - 16; Sofia: BAS, 1991, 20-33 (In Bulgarian).

19. Kotsilkova R., Ivanov Ya. "Rheology of fiber and particle reinforced polymers." In *Theoretical and Applied Rheology*, vol. 2, P. Moldenaers and R. Kennings, eds. Amsterdam: Elsevier, 1991, 859-62.
20. Ivanov Ya. "Effect of polydispersity on the flow behaviour and properties of concentrated suspensions." In *Advance in Structured and Heterogeneous Continua*, D. A. Siginer and Yu. Yanovsky, eds. N. Y.: Allerton Press, 1995, 35-8.
21. Ivanov Ya. "Microrheology of polymer composite with polydisperse solid phase." In *Progress and Trends in Rheology IV*, C. Gallegos, ed. Darmstadt: Steinkopff, 1994, 54-6.
22. Ivanov Ya. Structure and rheology of concentrated systems. Proceedings of the International Conference on Mechanics and Technology of Composite Materials; 1997 September 29 - Oktober 2; Sofia: BAS, 1997, 242-8 (In Bulgarian).
23. Yanovsky, Yuri, *Polymer Rheology: Theory and Practice*. London: Champan and Hall, 1993.
24. Sherman, Philip, *Industrial Rheology*. London: Acad. Press, 1970.
25. Nielsen, Laurence, *Mechanical Properties of Polymers and Composites*, vol. 2. N. Y.: Marcel Dekker, 1974.
26. Han, Chang, *Rheology in Polymer Processing*. N. Y.: Acad. Press, 1976.
27. Nielsen, Laurence, *Polymer Rheology*. N. Y.: Marcel Dekker, 1977.
28. Voznesenski, Vitalii, Lyashenko, Tatiana, Ivanov, Yatchko, Nikolov, Ivan, *Computer Aided Optimization of Composite Materials*. Kiev: Budivelnik, 1989 (In Russian).
29. Moshev, Viktor, Ivanov, Vasilii, *Rheological Behaviour of Concentrated Newtonian Suspensions*. Moscow: Nauka, 1990 (In Russian).
30. Shenoy A. "Rheology of highly filled polymer melt systems." In *Encyclopedia of Fluid Mechanics*, vol. 7, N. Chrenisinoff, ed. Huston: Gulf Publ., 1988.
31. Nakajima N., Harrel E. "Rheological behaviour of rubber carbon black compounds." In *Encyclopedia of Fluid Mechanics*, vol. 9, N. Chrenisinoff, ed. Huston: Gulf Publ., 1990.
32. Yanovsky Yu., Zaikov G. "Rheological properties of filled polymers." In *Encyclopedia of Fluid Mechanics*, vol 9, N. Chrenisinoff, ed. Houston: Gulf Publ., 1990.
33. Roskoe R. The viscosity of suspensions of rigid spheres. Brit. J. Appl. Phys. 1952; **3**: 267-79.
34. Maron S., Pierse P. Application of Ree - Eyring generalized flow theory to suspensions of spherical particals. J. Colloid Sci. 1956; **11**: 80-95.
35. Chong J., Christiansen E., Baer A. Rheology of consentrated suspensions. J. Appl. Polym. Sci. 1971; **15**: 2007-21.
36. Lee D. The viscosity of concentrated suspensions. J. Appl. Polym. Sci. 1969; **13**: 273-88.
37. Brady J., Vicic M. Normal stress in colloidal dispersions. J. Rheology 1995; **39**: 545-66.
38. Ivanov Ya., Kotsilkova R. On the viscosity of highly filled thermosetting composites. Proceedings of IUPAC-MACRO'83; 1983 September 5 - 9; Bucharest, 1983 (In Russian).
39. Ghosh A., Bhattacharya S. Rheological study of black coal-oil suspensions. J. Rheology 1984; **23**: 195-206.
40. Feodors R. Relationship between viscosity and concentration for Newtonian suspensions. J. Interf. Colloid. Sci. 1971; **46**: 545-7.
41. Graham A., Steele R., Bird R. Particle in concentrated suspensions: I. Experimental observation of particle clusters; II. Information theory and particle clusters; III. Prediction of suspension viscositty. Ind. Eng. Chem. Foun. 1986; **23**: 406-25.
42. Rutgers R. Relative viscosity of suspensions of rigid sphers in Newtonian liquids. Rheol. Acta 1962; **2**: 202-10.
43. Gillespie Th. The effect of aggregation and particle size distribution on the viscosity of Newtonian suspensions. J. Colloid Interface Sci. 1983; **94**: 166-73.
44. Barnea E., Mizzahi J. Dynamics of colloidal dispersions, J. Eng. Chem. Sci. 1973; **5**: 171-89.

45. Leighton O., Acrivos A. The shear-indused migration of particale in concentrated suspensions. J. Fluid Mech. 1987; **181**: 415-39.
46. Yanovsky Yu., Vinogradov G., Barancheyeva V. Rheological properties of compositions prepared from reclaimed thermoplastics. Vysokomol. Soedin. 1986; **5**: 983-900 (In Russian).
47. Aranguren M., Mora E., De Groot Jr.J., Macosko C. Effect of reinforcing fillers on the rheology of polymer melts. J. Rheology 1992; **6**: 1165-82.
48. Bretas R., Powell R. Dynamic and transient rheological properties of glass filled polymer melts. Rheol. Acta 1985; **24**: 69-84.
49. Metzner A. Rheology of suspensions in polymeric liquids. J. Rheology 1985; **29**: 739-75.
50. De Groot V., Macosko C. "Dynamic properties of model filled polymer melt in the linear viscoelastic region." In *Theoretical and Applied Rheology*, vol. 1, P. Moldenaers and R. Kennings, eds. Amsterdam: Elsevier, 1992, 339-41.
51. Marti-Martinez J., Macia-Agullo T., Fernandez-Garcia J., Ordiles-Barcelo A., Torro-Palau A. Properties of solvent based polyurethane adhesives containing fumed silicas. Macrom. Symp. 1996; **108**: 269-78.
52. *Polymer Composites*, Sedlacek, Blahoslav, ed. Berlin: Walter de Gruyter, 1986.
53. Friedrich Ch., Dehno H. "Viscoelastisity of silica and glass bed filled melts." In *Progress and Trends in Rheology IV*, C. Gallegos, ed. Darmstadt: Steinkopff, 1994, 45-7.
54. Clement F., Bakobza L., Monnerie L., Valet J. Analyse rheologique de reseaux de poly(methylsiloxane) sharges a la silice. Les Cahiers du Rheologie 1998; **16**: 203-8.
55. Heymann L., Aksel N. "Influence of surface modified particles on the rheological properties of suspensions." In *Progress and Trends in Rheology*, Igor Emri, ed. Darmstadt: Springer, 1998, 560-1.
56. Ishida H., Jang J. Role of the interface in multicomponent materials. European Symposium on Polymer Materials; 1987 September 14 - 18, Lyon, 1987, 1-12.
57. Han Chang, *Multiphase Flow in Polymer Processing*, N. Y.: Acad. Press, 1981.
58. Uriev, Naum, Ivanov, Yatchko. *Structure Formation and Rheology of Inorganic Disperse Systems and Materials*. Sofia: BAS, 1991 (In Russian).
59. Pahovichishin S., Tretinnik V., Gritsenko V., Shimanski A. Rheological properties of the dispersion pirogenic oxids of aluminium and silice in polymethylsiloxane. Coll. J. 1999; **2**: 247-53.
60. Kruglitskaya, Valeria, Zavadskaya, Zinaida, *Interfacial Phenomena in the Disperse Systems*. Kiev: Naukova Dumka, 1971 (In Russian).
61. Kruglitskii, Nikolai, Kruglitskaya, Valeria, *Disperse Structure in Organic and Silan - Organic Mediums*. Kiev: Naukova Dumka, 1984 (In Russian).
62. Koshevar V. Rheology of highly concentrated dispersions of ZnO particles in polymer containing mediums, including Lewis acids and basics, Coll. J. 1999; **2**: 215-21.
63. Trompette J., Grassl B., Fabregue E., Charnay C., Partyka S. Stabilite cinetique et comportement rheologique des suspension gelefiees talc-polyacryl glycinamide. Les Cahiers du Rheologie 1998; **16**: 52-8.
64. Baylocq P., Cartalos U., Piau J. M. "Rheological characterisation and structural modelling of clay polymer system." In *Progress and Trends in Rheology IV*, C. Gallegos, ed. Darmstadt: Steinkopff, 1994, 618-20.
65. Hesslink T. "Adsorption from solution at the solid/liquid interface." In *Parfitt CD*, C. H. Rochester, ed. London: Acad. Press, 1983, 64-87.
66. El Bounia N.-E., Brochette P. Effect of treating fillers by surfactants on the rheological behaviour of dispersions. Les Caiers de Rheologie 1998; **16**: 11-18.
67. *Chemistry and Technology of High Molecular Compositions* , vol. 14, A. Askadskii, ed. Moscow: VINITI, 1988 (In Russian).

68. Simeonov, Jordan, Hristova, Yulia, *Polymer Conctete. Structure, Properties and Mechanical Behaviour of Polymer Concrete*. Sofia: BAS, 1980 (In Bulgarian).
69. Arefiev, Boris, *Physico-Chemical Basis of the Compacting of Fiber Filled Composites Materials*. Moscow: Metalurgia, 1988 (In Russian).
70. Tsai, Shirley, *Composites Design*, (4th ed.), Paris, Dayton: Think Komposite, 1988.
71. Privalko, Vadim, Novikov, Viktor, Yanovsky, Yuri, *Basis of the Thermodynamics and Rheophysics of Polymer Materials*. Kiev: Naukova Dumka, 1999 (In Russian).
72. Lavrischev L., Trapeznikov A. Effect of aerosil on the propertis of polyester resin. Plast. Massy. 1969; **1**: 52-5 (In Russian).
73. Pugachevskaya N., Kovriga V. "Filled composition from nonsaturated polyesters." In *Chemistry and Technology of High Molecular Compositions*, Moscow: VINITI, 1981, 3-78 (In Russian).
74. Alt B. The thickening of unsaturated polyester resine with metal oxides. Kuntstoffe 1976; **12**: 786-92.
75. Burns R., Gandhi K., Hankin A., Lynskey B. Variability in sheet molding compound (SMC): I. The thickening reaction and effect of raw materials. Plastics and Polymers 1975; **43**: 228-35.
76. Ivanov Ya., Voznesenski V., Lyashenko T. "On the modelling of rheological properties of concentrated disperse systems." In *Advance in Rheology*, vol. 4. Mexico: Application, 1984, 517-22.
77. Ivanov Ya., Voznesenski V., Lyashenko T. "The use of experimental-statistical models in rheology of composites." In *Mathematika* vol. 9, "Metody matematiczne w technice", t. II, Opole, 1985, 165-76 .
78. *Fillers in Poymer Composite Materials*, G. Kats and D. Milevski, eds. Moscow: Chimia, 1981 (In Russian).
79. Wake, William, *Fillers in Composite Materials*. London: ILIFFE, 1971.
80. Paturoev, Vladimir, *Polymer Concrete*, Moscow: Stroiizdat, 1987 (In Russian).
81. Gudev, Nikolai, *Polymer Concrete, Polymer Mortars and Self Flowing Composites*, Sofia: Tehnika, 1981 (In Bulgarian).
82. Solomatov V., Knippenberg A. Investigations of the structure and properties of polyester polymer concrete. Izvestia Vusov 1977; **6**: 68-73 (In Russian).
83. Voznesenski V., Lyashenko T., Ivanov Ya. "Influence of the shear rate on the viscosity of a technological mixture." In *Investigations of the Properties of the Cement and Polymer Concretes*, Omsk: OPI, 1984, 10-18 (In Russian).
84. Bartenev, Georgi, Zelenev, Yuri, *Physics and Mechanics of Polymers*, Moscow: Visha Shkola, 1983 (In Russian).
85. Lipatov, Yuri, *Colloidal Chemistry of Polymers*, Kiev: Naukova Dumka, 1984 (In Russian).

Chapter 4

FORMATION OF POLYMER COMPOSITE MATERIALS IN THE PRESENCE OF SOLID ACIDS AND BASES

As noted in the previous chapters, the formation of polymers in the presence of acidic and basic fillers is determined by the adsorption character of the acid-base interaction within the compositions. When a filler - a solid acid or base, is introduced into a ready prepared polymer (basic or acidic, respectively), the polymer structure and properties may vary as a result of the adsorption interaction. These fillers affect the polymer more significantly when they are introduced during the stage of polymer formation, i.e. before the accomplishment of polymerization or polycondensation. Thus, they modify not only polymer structure and properties but also kinetics, thermodynamics and polymerization mechanism. The numerous published data show that solid acids and bases behave as effective catalysts of polymerization and polycondensation.

For the mechanochemically initiated polymerization of olefins in the presence of aluminium oxide [1] and in the presence of chromium oxide [2-4] it has been established that the reaction rate increases owing to the acidic sites that are located on the inorganic surface. The acceleration of styrene and butadiene polymerization in the presence of bentonite is established in [5]. Fukano and Kageyama have studied in detail the process of radiation-initiated polymerization of styrene in the presence of strong and weak solid acids, where aluminium oxide is used as a strong acid and aerosil - as a weak acid [6, 7]. Polymerization rate increase is observed in both cases and the effect is weaker when aerosil is present. It is found that different hydrated modifications of calcium sulphate act as catalysts during the polymerization of acrylamide and methyl methacrylate [8-10]. This is due to the presence of Lewis acid sites on the sulphate surface.

In practice, various clay minerals are widely used as fillers for polymers. As shown, they possess clearly expressed donor-acceptor abilities when interacting with various basic adsorbates and significantly stimulate polymerization processes taking place on its surfaces. It is known, for instance, that styrene polymerization considerably accelerates on the surface of montmorillonite. Polymerization rate is proportional to the number of the active acidic sites [11] where the latter comprise hydrogen and aluminium ions. Such a rate increase is established during the polymerization of isobutylene on montmorillonite, i.e. in the presence of an inorganic surface [12]. The favourable effect of the acidic inorganic surface is explained by the occurring donor-acceptor interaction between the polymer and filler. It helps the formation of a carbonium ion and a better ordering of the monomer molecules within the adsorption layer, thus relieving intermolecular interaction.

The acceleration of cation polymerization of styrene and isoprene on kaolinite is observed and reported in [13]. It is shown that the surface activation capability is completely suppressed when acid sites are neutralized by amines. Such an active site "poisoning" proves how important is their part in the acceleration of isoprene polymerization. Simultaneously, it is worth noting that polymer-filler adsorption interaction of an acid-base type may sometimes lead to polymerization inhibition. Thus, [14-16] show that the formation of a strong adsorption bond between the monomer molecules and the acidic filler surface yields a sharp decrease in the polymerization rate. Such an effect is observed in a 4-vinylpyridin/aerosil system. It is explained by the nature of the formed strong bonding, i.e. with the formation of two-centered complexes where carbonyl oxygen and the nitrogen atom of the lactam ring interact with the surface acidic hydroxyl groups. Polushkin and coauthors [14], prove that isoprene cationic polymerization on kaolin takes place when acid sites, located on the inorganic surface, participate in the reaction and polymerization is accomplished by protonization of the carbon double bond. The authors also prove that the quantity of formed polyisoprene depends on the number of acidic sites located on the kaolin surface. Many papers, for instance [15], reveal that organosilicon monomers are quite sensitive to the acidic nature of the contacting inorganic template. Such a sensitivity is registered during monomer adsorption and further thermal polycondensation. The studies establish that polymerization depends on the type of the surface acid sites - either protonic or aprotic. Moreover, when diphenylsilanediol adsorbs on titanium dioxide and aluminium oxide, its phenyl groups interact with Lewis and Brønsted inorganic surface acid sites. As a result, the Si-C_6H_5 bonds in the monomer molecule grows weaker, thus relieving thermal monomer polycondensation. Furthermore, the authors show that polysiloxanes formed in the presence of titanium and aluminium oxides possess a large number of branched molecules, while those which are formed in the presence of a weaker solid acid (copper oxide), have a linear structure with less branches. This means that polycondensation could be somewhat selectively affected by the solid acids. The authors also establish that a heat acidity-increasing treatment of aluminium oxide up to 400 °C leads to an increase of its activity during the thermal polycondensation.

The ability of some solid acids to affect selectively the process of polymerization is illustrated in the works of Natta and coauthors [16, 17]. They establish that poly-ketone may be obtained in the presence of strong acids of Lewis type (aluminium chloride and bromide). This could be a result of the strong adsorption interaction of the monomer (dimethylketene) with the salt acid sites. Besides, catalysts of basic and acidic character initiate the polymerization of diphenylketene in a different manner and yield the formation of polymers of various chemical structure. When polymerization takes place in the presence of tin tetrachloride, having acid sites on its surface, a polymer with a polyketene structure is formed. Upon substitution of the tin salt for basic-in-character lithium compounds, the polymerization yields a polymer with different structure, i.e. an unsaturated polymer.

[18] establishes and explains from an acid-base point of view the selective catalytic effect of barium sulphate on the possible directions of polycondensation of aminoaldehyde oligomers. Besides, the study reveals that barium sulphate doubles the polycondensation rate, and decreases polycondensation activation energy from 72 to 46 kJ/mol.

The selective catalytic performance of solid acids and bases during polymerization and polycondensation processes can be linked with the well known difference between homogeneous and heterogeneous catalysis. This difference consists mainly in formation of a specific intermediate compound proceeding during a heterogeneous catalysis in-between all reaction-capable compounds that can be formed as a result of the adsorbate-adsorbent interaction. Hence, polymerization or polycondensation in the presence of solid acids or bases takes place principally along one of all the possible directions.

Lipatova and coauthors examined the structure and properties of interlaced polyurethanes that are formed in the presence of aerosil [19, 20]. They showed that the compound structure and properties are determined principally by the acid-base interaction of isocianates with the active aerosil surface sites. The authors emphasise that the aerosil acidic surface acts as a catalyst during the formation of polyurethane. [19], in particular, treats the problems of formation of a linear polyurethane in the presence of aerosil. As noted there, the catalytic effect of the inorganic surface on the polymer formation from isocianate and polyoxypropyleneglycol can be linked with their acid-base interaction. Thus, the hydroxyl OH-bond grows weaker, a proton could be broken up, and its transfer to nitrogen in the course of urethane grouping formation facilitated.

As illustrated in Table 1, the introduction of 20 % aerosil (by weight) leads to 2 - 4 times increase of the polymerization rate constant and a halving of the activation energy. At the same time, aerosil modified by surface adsorption of decyl alcohol does not play an active part in this reaction. One of the possible reasons for the aerosil catalytic effect over the kinetics of polyurethane formation may be the polymer adsorption interaction with the aerosil acidic hydroxyl groups that facilitates urethane bond formation. After blocking the hydroxyl groups by introducing decyl alcohol, the filler becomes inactive.

Table 1. Kinetic parameters of the interaction between toluene-2,4-diisocianate and polyoxypropyleneglycol in filled and unfilled systems

Filler	Filling degree, % by weight	*T*, K	*K*.10^3, 1000 g/gequiv.min	$E_0 \pm 0.5$, kcal/mol
-	0	350	12.0	10.0
-	0	303	1.5	-
Aerosil	20	350	27.0	5.0
	20	303	7.0	-
Modified aerosil	20	350	11.5	-

The study of olefine oxides polymerization on the surface of aluminium and ferric oxides also proves that polymerization accelerates due to opening of the rings located at the oxygen atom. This occurs after the interaction of the oxygen unshared electron pair with the coordinatively unsaturated metal atom [21]. We should also note that the treatment of the oxide surface with protonic acids leads to their decreased activity in the course of polymerization of ethylene and propylene oxides. The authors of the study assume that the surface activity suppression after a subsequent treatment is related to screening of the active sites of aprotic acidic type - those are surface coordinatively unsaturated atoms of aluminium and iron. When another adsorbate is used (for instance, ehtyleneamine), the oxide protonic sites, i.e. protonized hydroxyl groups, turn out to be active throughout its polymerization. Hence, the catalytic character of one or another type of surface acidity is determined by the nature of the silicon dioxide which is used. In these cases the surface sites may serve as accelerators of the polymerization of acrylates and other similar monomers. This arises from the monomer donor-acceptor interaction with the inorganic surface; the catalytic ability of the surface is assigned to its acidic nature. This surface character, however, provides fast degradation of the initiator that governs polymerization and is in charge of monomer molecule orientation and ordering on the solid surface. Employing infrared spectroscopy, the authors of [22] establish that if a radiation polymerization of vinyl acetate develops on aerosil, then surface hydroxyl groups of acidic nature interact with the monomer carbonyl group, and this process affects the reaction rate. Note that an increase of the polymerization rate, as compared to that of a unfilled polymer, is established during the formation of polyester composites filled with quartz [23]. It is supposed that this phenomenon is due to the specific donor-acceptor interaction between the basic-in-character oligomer carbonyl group and the Brønsted acid sites located on the quartz surface. As a result, structures ordered on a supermolecular level are formed close to the filler particles. This conclusion is confirmed by data obtained after quartz treatment with a basic-in-character compound - octadecylamine. Interacting with the surface acid sites of quartz, octadecylamine blocks or deactivates these acid sites. Such a site deactivation yields a lack of ordered structures next the filler surface and a decrease in the polymerization rate. Similar effects are also registered when another acidic-in-character filler - titanium dioxide is added. A possibility for the development of an acid-base interaction between the oligomer carbonyl groups and coördinatively unsaturated Ti^{4+} ions is outlined. The increase of the polymerization rate of acrylic monomers in the presence of aerosil is referred to the formation of ordered structures that come out as a result of donor-acceptor interactions [24].

The results submitted in [25, 26] are also of interest. The authors study how mineral salts affect the structure of thiocol. They show that nitrates, sulphates and chlorides of some metals that behave as solid acids of a Lewis type, cause rubber structurizing. Moreover, using infrared spectroscopy and performing rheological studies, the authors conclude that the structurizing is due to a complex formation between coordinatively unsaturated cations of salt and rubber electron-donor groups -SH, due to the unshared electron pair of sulfur atoms.

The study of polymerization of epoxyacrylic oligomers in the presence of some metal oxides reveals that chromium oxide accelerates polymerization due to the interaction of its surface aprotic sites (i.e. coordinatively unsaturated chromium atoms) with the oligomer epoxy group [27]. Using mixed copper and cobalt silicates as fillers of polyester oligomers, the authors show that an acceleration of the polycondensation develops in this case [28]. The authors relate this phenomenon to an acid-base interaction of the cobalt and copper coordinatively unsaturated atoms with the oligomer basic-in-character groups.

In contrast to a process that develops under homogenous conditions, the polymerization of aminoacids in the presence of kaolinite and montmorillonite yields an increase of polymer production and polymerization degree. Such an effect is registered in [29] and the authors assign it to the interaction of a strongly ionized carboxyl group with the coordinatively unsaturated Al^{3+} ions, i.e. with a surface acid site of an aprotic type. It is also shown that in this process activation role can additionally play the exchanged cations, as well as the occurring protonic transfer from the surface towards the aminogroups of the aminoacid molecules. In their turn all these phenomena yield a polymerization activation.

When polymerization of methyl methacrylate takes place as initiated by benzoil peroxide, the introduction of titanium dioxide or clays allows for an acceleration of the process and for a significant increase of its activation energy [30-33]. The authors link the observed effect with an adsorption interaction of acid-base character that takes place between the surface active sites of the titanium dioxide and the monomer basic-in-character groups. In this case the hydroxyl groups are referred to as active surface acid sites.

[34] treats the catalytic effect of some metal chlorides over the polymerization of methyl methacrylate in the presence of an initiator - 2,2'-azobisisobutironitrile. Such an effect is determined by the formation of a complex compound with an increased reaction ability, because of the interaction between the basic-in-character carbonyl groups and the surface Lewis acidic sites of zinc, copper, cobalt, and nickel chlorides. The increased reaction ability of the complex-forming monomer is explained by the delocalization of electrons in its double bond:

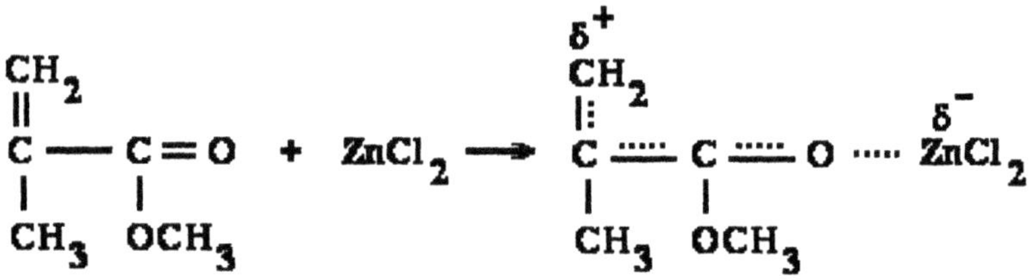

Chlorides not only accelerate the methyl methacrylate polymerization but significantly increase the degree of monomer conversion. The infrared data for those systems show that this kind of complexation shifts the $\nu_{C=O}$ band at about 100-150 cm^{-1}. The same paper shows that the activation energy of the polymerization of methyl methacrylate in the presence of $ZnCl_2$ is half that of a homogeneous polymerization.

An accelerating effect of lead acetate on the polymerization of caprolactam is established in [35] and is illustrated in Fig. 1.

This lead acetate behaviour is explained by the interaction of its surface aprotic sites (coordinatively unsaturated lead cations) with the carbonyl oxygen atom of the lactam amide group. As a result, the nucleophilicity of the amide group and the acilation ability of the coordinated monomer molecule strongly increase. This aids the formation of N-aminocaproylcaprolactam that itself stands for a site of polymer chains growth.

[36-40] present a study of special features of the lactam polymerization in the presence of solid acids - oxides of transition metals. Such are titanium, molybdenum, chromium, cobalt and some other oxides. It is established that these compounds prove to be effective catalysts during the ε-caprolactam polymerization. They accelerate the polymerization process and decrease the effective activation energy. A comparison with the process of ε-caprolactam autopolymerization is made. It is supposed that the oxide acidic sites are able to activate monomer molecules, and polymerization initiation is linked with an electron transfer from the monomer amide group to the coordinatively unsaturated metal cations. Hence, ionization of the adsorbed molecules of ε-

caprolactam takes place on the oxide surface, whereat the adsorbed molecules have a decreased resonance stability due to the violation of the conjunction between the unshared electron pair of the nitrogen atom and the π-bond of the carbonyl group. This yields weakening of the amide bond and leads to a decrease of the effective activation energy. It is also established that from among oxides the most effective catalyst of polymerization is the titanium dioxide, characterised by strongly expressed electron-accepting properties, i.e. it proves to be a strongest acid of Lewis type. In addition the authors show that compounds with basic-in-character surface properties (in this case calcium carbonate) are totally inert throughout ε-caprolactam polymerization.

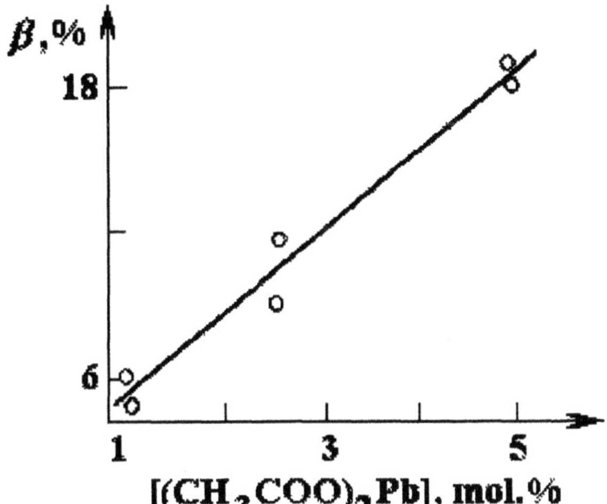

Figure 1. Polycaproamide yield *vs.* lead acetate concentration at 255 °C and for polymerization time 4 hrs

Table 2. Comparative data for autopolymerization of caprolactam (CL) and dodecalactam (DDL) and for heterogeneous polymerization at 573 K

Monomer	Initiator 5 %	Conversion rate, %/h	Induction period	E_{eff}, kJ/mol
DDL	-	1.5	-	255
	TiO_2	42	-	129
	Cr_2O_3	9.0	-	129
	ZnO	7.0	-	129
CL	-	7.4	10	121
	TiO_2	21.0	0.8	88

Table 2 presents values of the reaction rate for an initial stage of the DDL polymerization process, while CL data cover the maximal value of the reaction rate. In this case the initiating capability of the metal oxides, displayed during the lactam polymerization, is determined by Lewis acid sites located on their surfaces. These are coordinatively unsaturated metal ions that can accept electrons from the monomer amide group. Note that, as a result, an ionization of the lactam molecules can take place.

The lactam molecules thus ionized have a decreased resonance stability which is due to the violation of the conjunction between the unshared electron pair of the nitrogen atom and the π-double bond of the carbonyl group. In this way, a decrease of the amide π-double bond strength occurs which in turn helps to decrease of activation energy of lactam polymerization as compared to that of autopolymerization, and an increase of the process rate. Data for the polymerization of lactams in the presence of oxides of copper, vanadium and of some other metals prove that the lactam adsorption on metal oxides is accompanied by an electron transfer from the amide group to the surface of the solid acid, since these oxides are interaction-capable. In this case a reduction of the metal ions of oxides is really observed until attaining a lower valence.

We have studied ε-caprolactam polymerization in the presence of a number of solid acids - silicon and aluminium oxide, calcium phosphate, barium and calcium sulphates [41, 42]. The conditions of analysis of their acidic characteristics are given in Chapter 1, Table 1. We established that aluminium oxide and calcium phosphate are quite effective catalysts during the ε-caprolactam polymerization. We were able to accomplish the process without introducing other traditional initiators. Fig. 2 and Fig. 3 illustrate the kinetics of polycaprolactam formation in the presence of different portions of filler-catalysts in the monomer.

The figures show that the increase of the catalyst concentration, i.e. the number of acid sites in the system, leads to the monomer conversion rise. Such kinetic curves are shown in [39, 40], where ε-caprolactam polymerization in the presence of titanium dioxide is tackled.

Another noteworthy point is that polymerization in the presence of aluminium oxide and calcium phosphate runs practically without an induction period. This characteristic makes the process considerably different from autopolymerization which has a 50 hours induction period. Moreover, it is thus distinguished from ε-caprolactam hydrolytic polymerization - the latter is always preceded by an induction period which can vary widely in length depending on the water content of the system [41]. When the temperature is raised, a considerable increase of the rate of heterogeneous polymerization is observed, as seen in Fig. 4 and Fig. 5.

The overall activation energy of the polymerization process is calculated from the temperature dependences of the monomer conversion. When polymerization takes place in the presence of 20 wt. % of aluminium oxide, the activation energy is 66 kJ/mol, and in the presence of 20 wt. % of calcium phosphate it is 61 kJ/mol. For comparison, note that the value of the overall activation energy of the autocatalytic polymerization of ε-caprolactam exceeds 121 kJ/mol [36].

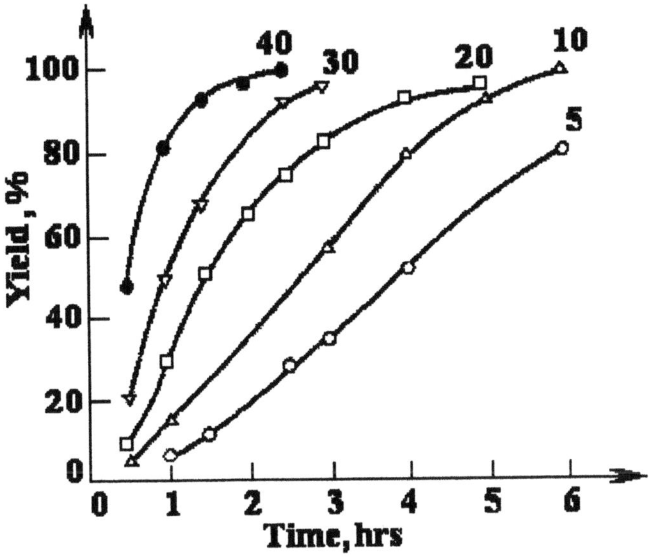

Figure 2. Polycaprolactam yield *vs.* polymerization time using variour concentrations of the filler-catalyst γ-Al_2O_3 at 523 K

Consider polymerization in the presence of γ-Al_2O_3. Then the rate constant values are $2.4.10^{-4}$ s^{-1}; $3.5.10^{-4}$ s^{-1}, and $6.6.10^{-4}$ s^{-1}, respectively, for 10, 20 and 40 wt. % oxide content and a temperature of 523 K. For calcium phosphate, the rate constant values are $1.8.10^{-4}$ s^{-1}; $3.1.10^{-4}$ s^{-1}, and $4.7.10^{-4}$ s^{-1} for salt content in the composition of 10, 20 and 30 wt. %. If we calculate the rate constant of autopolymerization from the data submitted in [42], we get only $1.9.10^{-6}$ s^{-1} for the same temperature. All these data illustrate the especially strong catalytic effect of aluminium oxide and calcium phosphate during the ε-caprolactam polymerization.

Consider now the specific surface of the solid acid. Then the values of the polymerization rate constants that characterize the specific catalytic activity of aluminium oxide and calcium phosphate are in a proportion 1 : 1.8 - the specific surface of the aluminium oxide is 155 m^2/g while that of calcium phosphate - to 82 m^2/g (data are obtained using the method of low-temperature adsorption of nitrogen). This means that, compared to aluminium oxide, calcium phosphate behaves as a more active catalyst during the ε-caprolactam polymerization. Such a phosphate characteristic can be explained by means of the heterogeneous-homogeneous mechanism of catalysis that we have supposed to function here. The basic stages of that mechanism consist in the following: an activation of the adsorbing molecules takes place as a result of the adsorption interaction between the monomer molecules and the active surface sites of the filler-catalyst. The monomer molecules thus activated desorb in the system volume where the polymerization itself takes place. Hence, on the one hand, the active surface sites should be strong enough to activate the monomer molecules, while on the other hand such a molecule-active site interaction should not be too strong as to hamper the activated monomer molecules desorption in the system volume. The smaller polymerization rate in the presence of aluminium oxide compared to that for calcium phosphate can be explained from that point of view. The higher acidity of the aluminium oxide causes formation of a stronger adsorption bond with the monomer molecules and reduces their desorption rate. Note also that, as a result, in the presence of aluminium oxide as compared to the calcium phosphate, the polymerization reaction itself slows down.

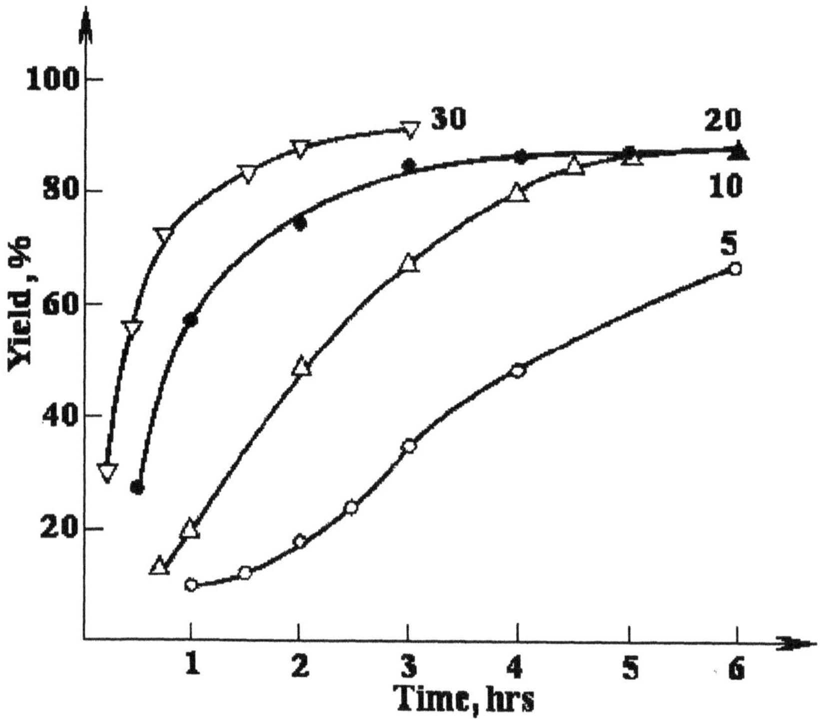

Figure 3. Polycaprolactam yield *vs*. polymerization time using variour concentrations of the filler-catalyst $Ca_3(PO_4)_2$ at 523 K

Conversion of the ε-caprolactam, polymerized in the presence of two types of calcium phosphate (see Table 3 inChapter 1), is shown in Fig. 6. It is seen that the difference in the acidity of the two types of calcium phosphate is revealed for a filling degree of up to 20 %, while the existence of weaker (but at a higher concentration) acidic sites leads to faster polymerization.

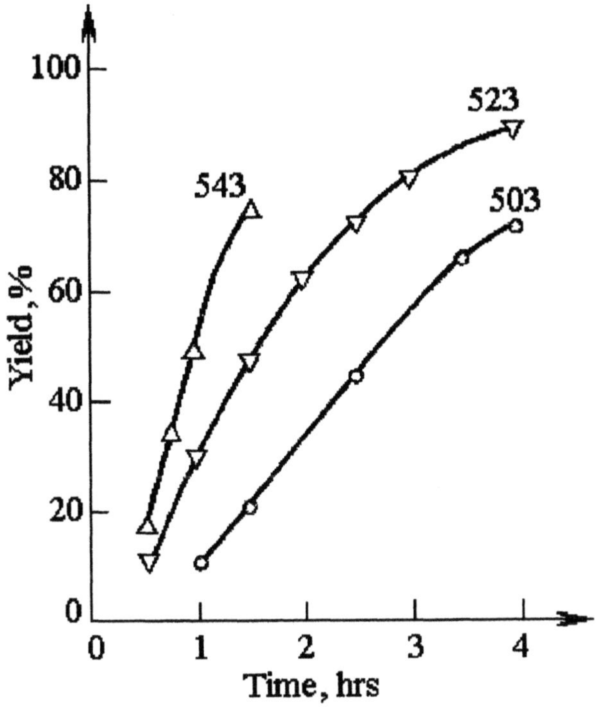

Figure 4. Polycaprolactam yield *vs.* polymerization time for 20 wt. % concentration of the filler-catalyst γ-Al_2O_3 at various temperatures

Regarding the two solid acids that act as catalysts, ε-caprolactam polymerization turns out to be a first order monomer reaction, since the kinetic data given in Fig. 2 and Fig. 3 lie on a straight line as shown in Fig. 7 and Fig. 8, in coordinate systems subsequently introduced.

A weak deviation from linear dependence is observed during the early stage of reaction, only, and in the presence of aluminium oxide. This can be explained by the unsteady state of the process.

Considering the rest of the characteristics of the heterogeneous ε-caprolactam polymerization, it seems worth noting the following: Figs. 9 and 10 show the thermograms of polycaprolactam obtained at 523 K in the presence of calcium phosphate. The figures show that the increase of the polymerization time for a constant catalyst concentration (Fig. 9) or the increase of the catalyst concentration for a constant polymerization time (Fig. 10) yields a change of the ratio between the two caprolactam modifications, i.e between the low-temperature metastable γ^*-crystal form and the high-temperature α-crystal form that is known to be thermodynamically the most stable one. Hence, the catalyst significantly affects the formation of polymer crystal structure.

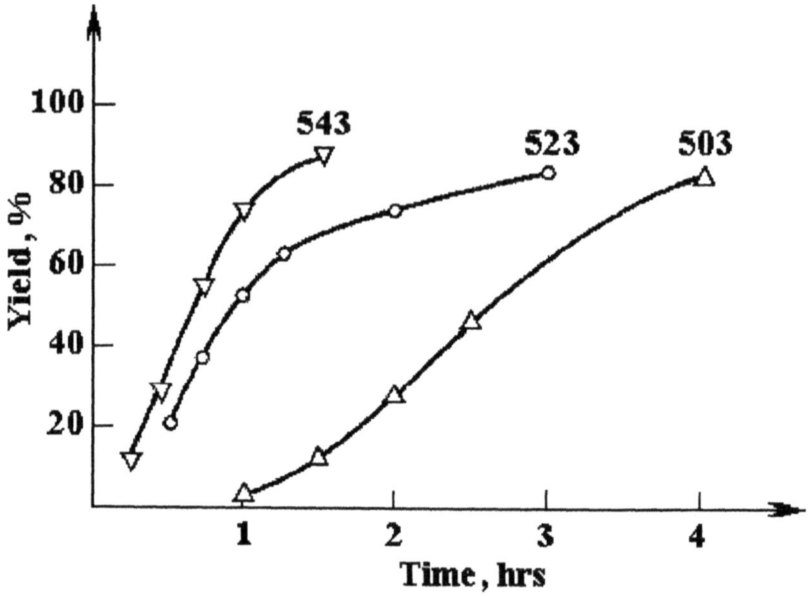

Figure 5. Polycaprolactam yield *vs.* polymerization time for 20 wt. % concentration of the filler-catalyst $Ca_3(PO_4)_2$ at various temperatures

From the shown thermograms we calculate the polycaprolactam degree of crystallinity using the melting heat of the obtained polymer as compared to the theoretical value 202.7 kJ/mol [42, 43], for a polymer with a 100 % degree of crystallinity. The results are given in Table 3.

Table 3. Degree of crystallinity of a polymerization-filled with calcium phosphate polycaprolactam versus the polymerization time and the concentration of the filler-catalyst at 523 K

[Filler] = 20 wt. %, *t* (hrs)					*t* = 3 hrs, [Filler] (wt. %)		
1	2	3	4	5	10	20	30
62	53	54	48	51	55	54	54

These data prove that a high degree of crystallinity is attained even during the initial period of synthesis (i.e. during the first hour), and is maintained both as polymerization proceeds and also with increase of filling for a given synthesis duration.

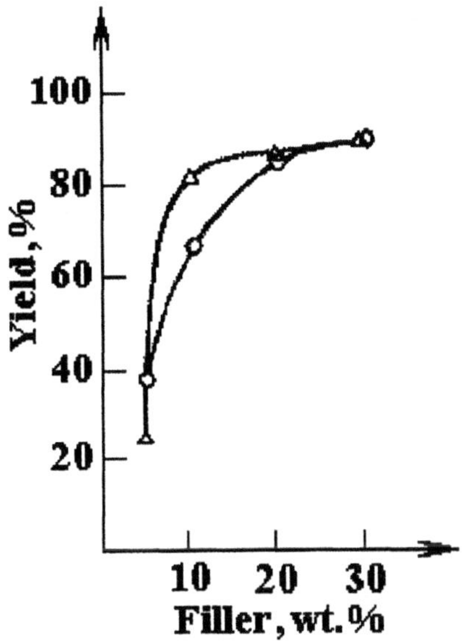

Figure 6. Conversion of polycaprolactam in the presence of two types of filler-catalysts: O - $Ca_3(PO_4)_2$ sort I ; Δ - $Ca_3(PO_4)_2$ sort II

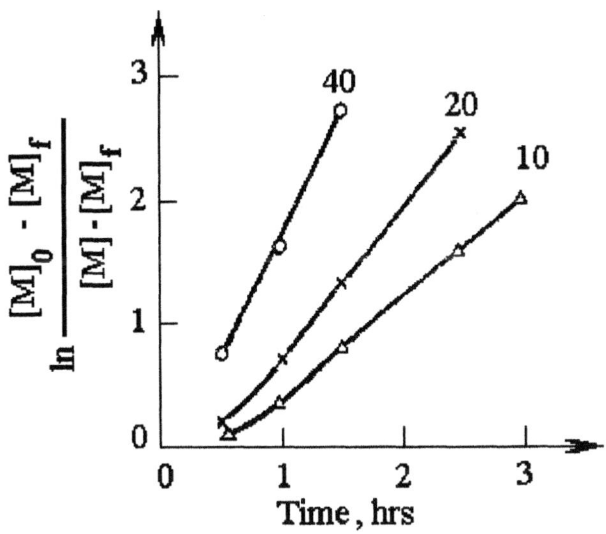

Figure 7. Logarithmic relative monomer concentration *vs.* polymerization time using various concentrations of the filler-catalyst γ-Al_2O_3 at 523 K: $[M]_0$ - initial monomer concentration; $[M]_f$ - final concentration; $[M]$ - current concentration

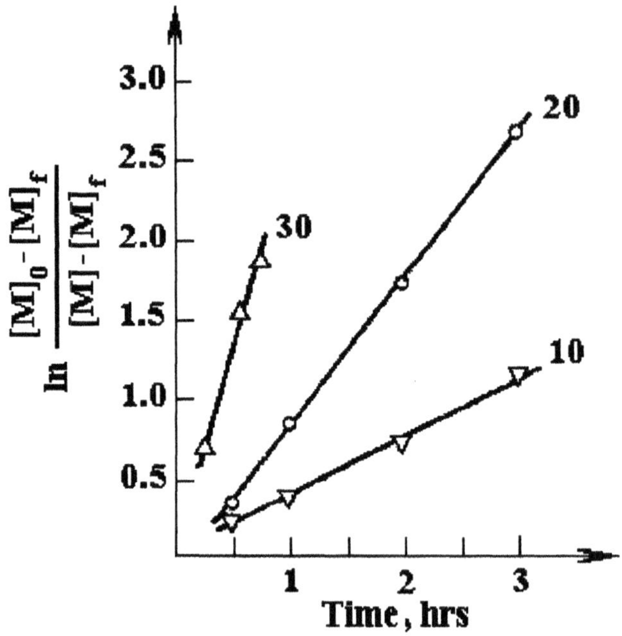

Figure 8. Logarithmic relative monomer concentration *vs.* polymerization time using various concentrations of the filler-catalyst $Ca_3(PO_4)_2$ at 523 K: $[M]_0$ - initial monomer concentration; $[M]_f$ - final concentration; $[M]$ - current concentration

When polymerization of ε-caprolactam takes place, its molecular weight also rises along with the monomer conversion, as seen in Table 4.

Table 4. ε-Caprolactam conversion and molecular weight of the polycaprolactam *vs.* time and temperature of polymerization for 20 % content of calcium phosphate

T, K	Time, hrs	Yield, %	$M_\eta \cdot 10^{-3}$
503	1.5	13	3
	2.5	47	5
	4.0	87	11
523	0.5	21	2
	1.0	54	5
	2.0	74	15
	3.0	84	21
	4.0	87	29
543	0.75	27	9
	1.25	76	18
	1.5	87	23
523*	3.0	87	43

Note: * Data are valid for 30 % content of calcium phosphate.

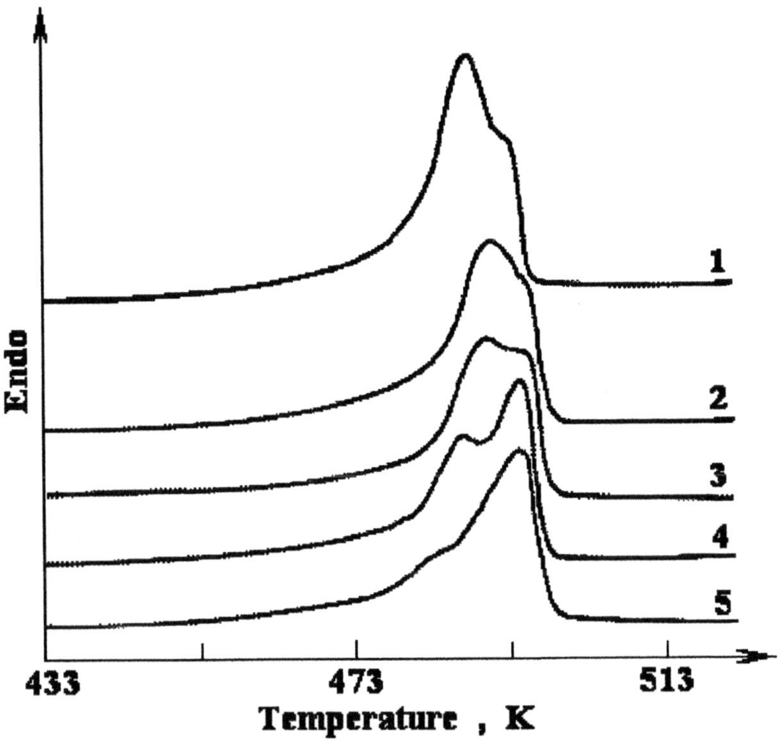

Figure 9. DSC-thermograms of polycaprolactam at 20 wt. % of $Ca_3(PO_4)_2$ for various times of polymerization: ***1*** - 1 hr; ***2*** - 2 hrs; ***3*** - 3 hrs; ***4*** - 4 hrs; ***5*** - 5 hrs

The temperature increase yields an increase of the polymerization rate and of the molecular weight during the initial stages of the process. At the final reaction stage, however, the largest molecular weight is registered at a temperature of 523 K and then the weight decreases. It is interesting that the molecular weight increases with the increase of the content of the solid acid in the system, for one and the same temperature and process duration. The data for the molecular weight of the polymer, prepared by using the method of polymerization filling, differ from those obtained by the traditional methods of ε-caprolactam polymerization. Thus, considering hydrolytic, anionic and cationic polymerization, the molecular weight does not rise with the temperature increase. Besides, under hydrolytic and cationic polymerization, the molecular weight of the formed polymer decreases with the increase of the initiator portion within the system.

Fig. 11 shows data on ε-caprolactam conversion *vs*. time as dependent on the acidic strength of the filler-catalysts for a constant concentration of the acid sites.

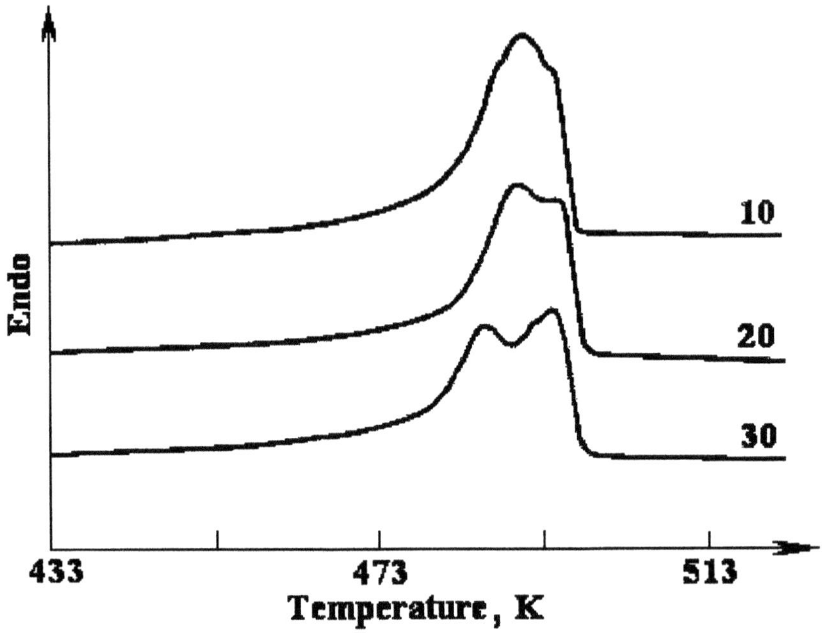

Figure 10. DSC-thermograms of polycaprolactam at various concentrations of $Ca_3(PO_4)_2$ (wt. %) for polymerization time 3 hrs

As is seen, with the increase of the strength of the acid sites from +6.8 to +2.0, the rate of ε-caprolactam conversion increases too, while its further increase (up to -3.0) is accompanied by a strong decrease of the catalyst activity. Hence, an optimum of the acidic site strength can be sought between +2.0 and +3.3 where the conversion rate is maximal - the equilibrium monomer concentration is attained for only two hours after starting the process. Based on the process scheme proposed, this effect can be explained as follows: catalyst acidic sites with strength lower than +3.3 are not able to weaken the amide bond in the monomer molecules much. Thus, it becomes harder to activate the molecules and the entire polymerization reaction slows down as a result. For acid sites of strength higher than +3.3, the reaction slows down due to the vigorous sites-monomer molecule interaction. This leads to harder desorption of the activated molecules from the surface into the volume, where other polymerization stages proceed. Hence, the existence of an optimum of the dependence between acid sites strength and their catalytic activity can be related to the specific characteristics of the adsorption interactions, determining the limiting stage of polymerization. If the acid site strength is kept constant and sites concentration is changed, i.e. by increasing the site number, dependences shown in Fig. 12 are obtained.

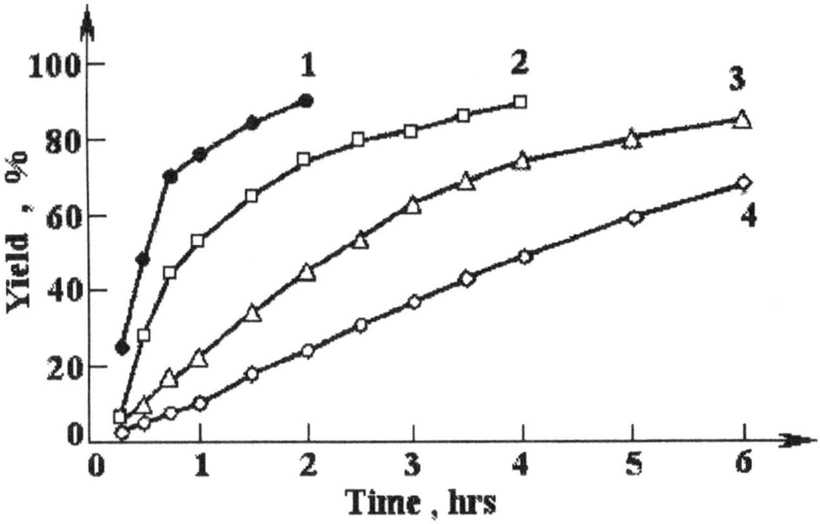

Figure 11. Caprolactam conversion vs. the acid sites strength for constant concentration of the acid sites 0.015 mequiv/g: ***1*** - pK_a = +2.0 - +3.3; ***2*** - pK_a = +3.3 - +4.8; ***3*** - pK_a = +4.8 - +6.8; ***4*** - pK_a = +1.5 - -3.0

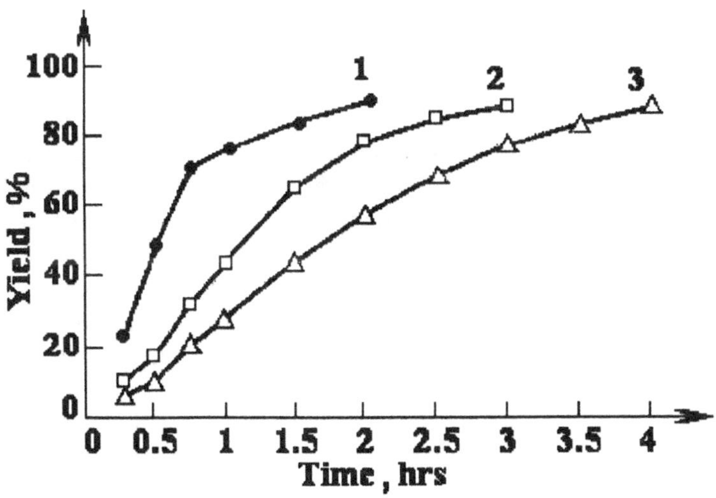

Figure 12. Caprolactam conversion *vs.* the acid sites concentration at constant acid sites strength pK_a = +3.3 - +4.8 (T = 523 K): ***1*** - 0.150 mequiv/g; ***2*** - 0.100 mequiv/g; ***3*** - 0.015 mequiv/g

The change of the concentration of the surface acidic sites is accomplished through a thermal treatment, as well as by preparing mixed catalysts (by mixing different proportions of aluminium oxide and calcium phosphate). When there are no acidic sites on the catalyst surface (for instance, thermally untreated aluminium oxide), ε-caprolactam polymerization practically does not take place if other initiators of the process are not introduced. All these data show that a definite correlation exists between the acidic properties of inorganic compounds and their catalytic performance during ε-caprolactam polymerization. Besides, the correlation does not depend on the type of catalyst used - aluminium oxide or calcium phosphate. Hence, it is possible to foresee the catalytic effect of the inorganic compounds during ε-caprolactam polymerization, based on data about their acid properties. To prove this assumption we chose and characterized solid acids - titanium and silicon dioxides, that have an acid strength of +4.8 - +6.8 and +3.3 - +4.8, respectively. It was established that both inorganic surfaces really catalyze ε-caprolactam polymerization. Besides, in these cases the rate of monomer conversion is practically the same as that when calcium phosphate of the same acid site strength is used as a catalyst. This proves that a prediction of the catalytic effect of the inorganic compounds during ε-caprolactam polymerization is possible. Moreover, the set of catalysts that could be applied for this process can be considerably widened in view of the large number and variety of existing solid acids.

The analyzed characteristics of the process of ε-caprolactam polymerization, catalyzed by solid acids, show that heterogeneous polymerization (or some of its separate stages) follows a mechanism that differs from homogeneous polymerization.

It is worth noting, however, that some of the parameters of those processes of different character are quite similar, e.g. (i) the activation energy of homogeneous and heterogeneous polymerization is almost one and the same; (ii) the rate of cationic polymerization during the initial stage of the process is close to that of heterogeneous polymerization [44, 45]. These facts can be considered as an indication that polymerization filling consists in separate stages that are similar to the known stages of hydrolytic and ionic polymerization. Hence, the study of heterogeneous polymerization systems can be based on a possible analogy with known conventional methods for polyamide synthesis - e.g. hydrolytic and cationic polymerization. Hence, based on these facts we performed a more detailed study of the kinetics of heterogeneous caprolactam polymerization in the presence of a catalyst - aluminium oxide.

The ε-caprolactam used throughout the experiments was recrystallized from toluene. It is further dried in vacuum at 300 K. This purified monomer does not produce any water-insoluble roducts when treated at 523 K for 20 hrs in the absence of initiation additives. The catalyst is γ-Al_2O_3, annealed at 523 K for 2 hrs, with a specific surface of 155 m^2/g. Its surface acidity is found throughout the experiment, too. Polymerization is carried out within a nitrogen flow in a laboratory reactor of stainless steel supplied with a screw stirrer and an electric heating system. The dispersed catalyst is added to ε-caprolactam - the latter is first heated to 523 K. Portions of 5 - 10 g from the reaction mixture are taken directly from the reactor after fixed consecutive time intervals and are subjected to analysis. The monomer content is determined by drying the unreacted ε-caprolactam for 4 - 6 hrs, in vacuo at 351 K. The process proceeds until getting ε-caprolactam of constant weight. Then the amounts of the end groups are determined by following the procedure outlined in [46]. Basic groups are titrated with 0.02 mol/l HCl in a solution of phenol + ethanol (70/30 vol. %). The end point is determined with an accuracy of 5 %. Acidic groups are titrated in hot benzyl alcohol with 0.02 mol/l KOH dissolved in benzyl alcohol - here the accuracy of the end point determination is 7 %. The catalyst - filler does not interfere with titration. The acid groups content coincides with that of the basic groups and the result is within the admissible error interval. The ε-caprolactam content and that of the end groups are corrected, following the change of the polymerizate density that takes place in the course of polyamidation. Data are submitted in mol per liter.

The presence of active acylating groups is checked by employing a qualitative test that produces coloured complexes with the hydroxamic acids. This is done by following the procedure given in [47]. The controlled cationic polymerizate is synthesized following [47]. Fig. 13 and Fig. 14 show the X monomer content and the Y end groups content plotted *vs*. time.

Note that plots are taken in the course of polymerization that takes place in the presence of 4, 8.8 and 15.5 wt. % of γ-Al_2O_3, respectively. A sharp increase of the end group content is observed in all the experiments during the first several minutes of the process. At the same time, the monomer content starts to drop. The end group concentration reaches a maximum and then slowly drops down. The maximum of end groups concentration value coincides with that of the maximum conversion rate.

The phenomena observed are discussed in view of the basic principles of designing hydrolytic [44] and cationic [45] models of polyamide formation. Three main reactions are considered during the polymerization of ε-caprolactam (CL) in the presence of water. These are chain initiation, chain growth and chain termination, schematically presented as follows:

$$CL + H_2O \Leftrightarrow H_2N{-}(CH_2)_5{-}COOH \quad (1)$$

$$P(n){-}NH_2 + CL \Leftrightarrow P(n+1){-}NH_2 \quad (2)$$

$$P(n){-}NH_2 + P(m){-}COOH \Leftrightarrow P(n+m) + H_2O \quad (3)$$

where *P(n)* and *P(m)* are polymers with single units *n* and *m*, respectively.

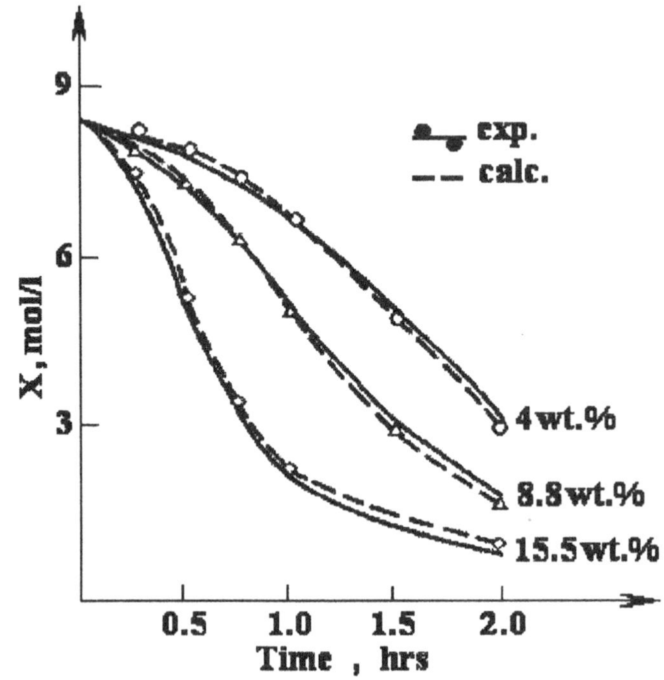

Figure 13. Calculated and experimental values of monomer canversion *vs.* polymerization time for various concentrations of the filler-catalyst

The start of cationic polymerization of the ε-caprolactam (e.g. by using HCl) is expressed by the reaction

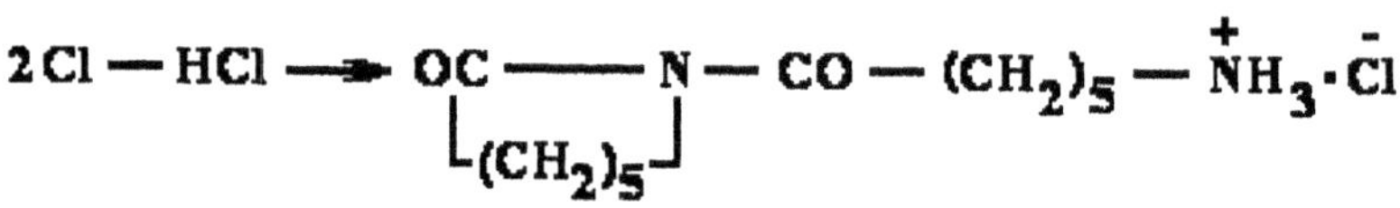

(4)

Note that reaction (4) and the subsequent polyaddition of ε-caprolactam run very fast. Moreover, the problem of how a preferred orientation of the ε-caprolactam is attained during its attachment is so far unsolved. The reaction is further retarded at this stage owing to the formation of amidine groups accompanied by the release of water molecules:

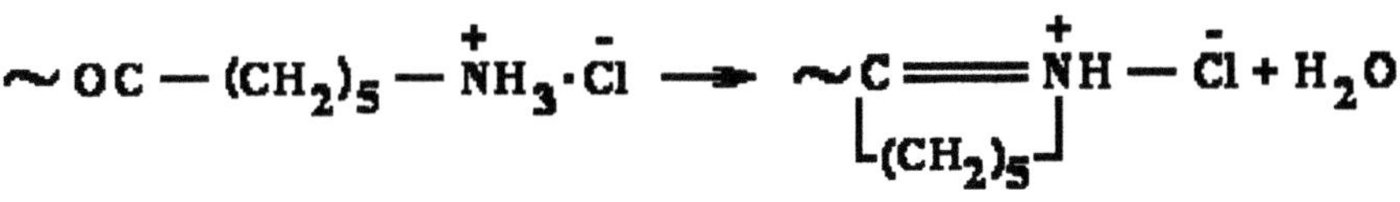

(5)

Reaction (5) sets the beginning of a relatively slow stage of the polyamidation reactions (1) - (3).

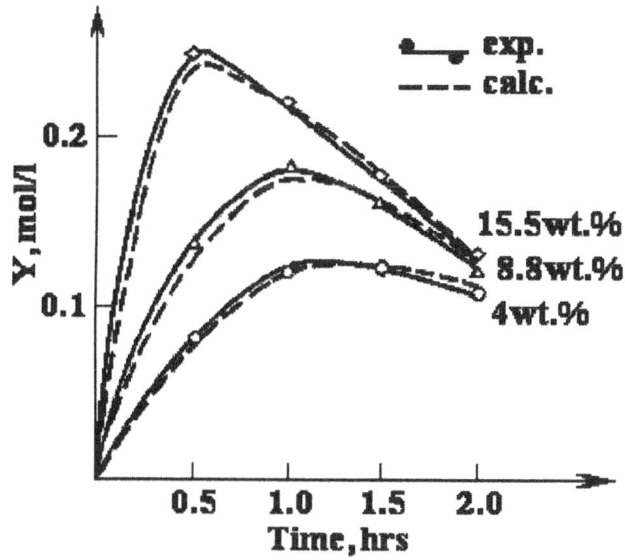

Figure 14. Calculated and experimental values of the content of end groups *vs.* for various concentrations of the filler-catalyst

Consider now in detail the stages of the studied heterogeneous reaction.

1. INITIATION STAGE. Following the hydrolytic model, end $-NH_2$ and $-COOH$ groups are formed due to the slow hydrolysis of the ε-caprolactam. The induction period takes almost one third of the polymerization time. In our case a sharp increase of the amount of the end groups is observed during the first minutes of the synthesis. Such a reaction course best fits the first stage of the cationic polymerization. We assume that the most probable reactions that take place during this stage are the fast ones running upon the Lewis acid sites of the γ-Al_2O_3 surface (i.e. upon the coordinatively unsaturated Al^{3+} ions). The reactions develop following the scheme

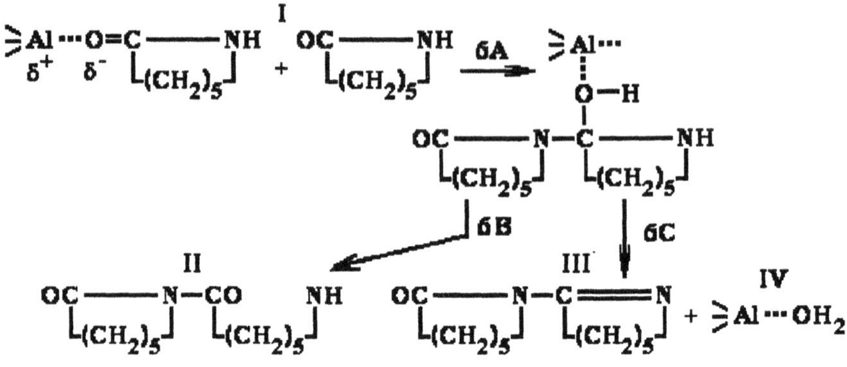

(6)

Note that reactions (6A) and (6B) are similar to those given in [45] that record the polymerization of ε-caprolactam in the presence of the soluble Lewis acids AlCl3 and TiCl4. The possible reaction (6C), being similar to reaction (5) that takes place throughout the initial polymerization stage, proves to be negative in contrast to the checked cationic polymerizate. Such a negative test result excludes the presence of detectable amounts of product II or of its analogues. On the other hand, aluminium oxide may provide a formation of products III and IV, owing to its dehydration ability [48]. Water molecules evolve on the surface acid site, following the scheme (6C) (product IV). They are able to quickly hydrolyze the ε-caprolactam, producing - NH_2 and -COOH groups that are needed for the chain growth. At that the active surface site is restored:

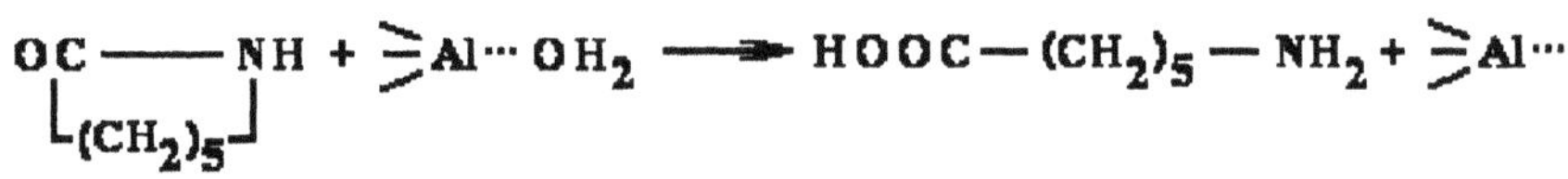

(7)

In fact, reactions (6) and (7) replace reaction (1). Their high rates may be referred to the adsorption interaction between the monomer and the surface acid sites of the filler-catalyst. This is confirmed by comparing the IR spectra of a condensed ε-caprolactam and that of ε-caprolactam adsorbed on the γ-Al_2O_3 surface. A shift of 45 cm^{-1} to the low frequency range of the carbonyl group is observed, confirming the occurrence of a strong acid-base interaction that results in the formation of an intermediate complex C=O...Al between the monomer and the surface Lewis acid sites. As with proton attachment [45], such a type of complex formation should sharply weaken the entire amide bond, due to the partial breaking of the π-bonding in the O=C-N group on the one hand, and should obviously facilitate the surface reactions of the ε-

caprolactam. It should be noted that even though the equations fit with literature and experimental data, the final confirmation requires further studies.

2. CHAIN GROWTH. There are several points confirming the suggestion that the chain growth runs following reaction (2), the latter being inherent to the hydrolytic polymerization: (i) a similarity established between hydrolytic and heterogeneous polymerization, i.e. lack of diffusion limitations on the chain growth, and good agreement between the values of activation energy and equilibrium monomer concentration [41, 42]; (ii) good agreement between the end groups content and monomer conversion rate (Figs. 13 and 14).

Consider now that reaction (2) is the main one of monomer consumption, as is assumed in a hydrolytic model. Then, the conversion rate can be given as $dX/d\tau$.

$$dX/d\tau = k_{21}XY + k_{22}Y \tag{8}$$

where k_{21} the reaction rate constant and indices *1* and *2* denote the direct and the opposite reaction, respectively.

3. CHAIN TERMINATION. The polycondensation reaction (3) involved in the hydrolytic model runs during a heterogeneous synthesis, too. In this case, however, the quantity of evolved water is rapidly evaporated in an open-type reactor due to the high temperature. Then, reaction (3) should be considered as an irreversible one. Based on equations (3) and (7), one can write the differential equation for the end groups in the following form:

$$dY/d\tau = k_{71}BX - k_{31}Y^2 \tag{9}$$

where $B = C.[OA]$ is the concentration of the active hydrating sites (IV) per unit volume of the liquid phase, $[OA]$ is the catalyst content expressed in the surface units per unit volume of the liquid phase, m^2/l, and C is the concentration of (IV) per surface unit. When modelling the kinetics of formation and loss of end groups, the following hypothesis is formulated:

a) concentration B is constant in the course of synthesis, i.e. reactions (6A) and (6C) are slower than reaction (7);
b) concentration B is proportional to the consumption of ε-caprolactam X, i.e. reactions (6A) and (6C) are slower than reaction (7). Then, equation (9) can be written in two forms:

$$dY/dt + k_{31}Y^2 = a_1X[OA] \tag{10A}$$

and

$$dY/d\tau + k_{31}Y^2 = a_2X^2[OA] \tag{10B}$$

where a_1 and a_2 are equal to the product $k_{71}.C$ and their values are calculated for each experimental point, the value of k_{31} is taken from [46] and the remaining quantities are obtained on the basis of experimental data.

In all the experiments, the parameter a1 appears to be the most stable one and its value varies within the ranges $2 \times 10^{-6} \div 3 \times 10^{-6}$ l/ m^2h. This confirms that the concentration of the active site (IV) is kept constant during polyamidation - that is the value *C*. Numerical integration of the kinetic equations (8) and (10A) is performed by using 4th order Runge-Kutta-Gill method and applying a step of 0.002 h. Figs. 13 and 14 show the *X* monomer content and the *Y* end groups content plotted *vs*. time. The results are plotted by dotted lines in Figs. 13 and 14, while the quantities k_{21}, k_{22} and k_{31} are taken from [46]. The good agreement between the experimental and calculated plots proves the validity of the proposed reaction scheme.

The study [47] shows that polymerization of epoxy oligomers significantly speeds up in the presence of mica. It is supposed that this effect is due to mica participation in opening of the monomer epoxy ring, resulting from the donor-acceptor interaction between mica surface hydroxyl groups and the oxygen atom of the epoxy ring. The authors propose the following scheme of the process:

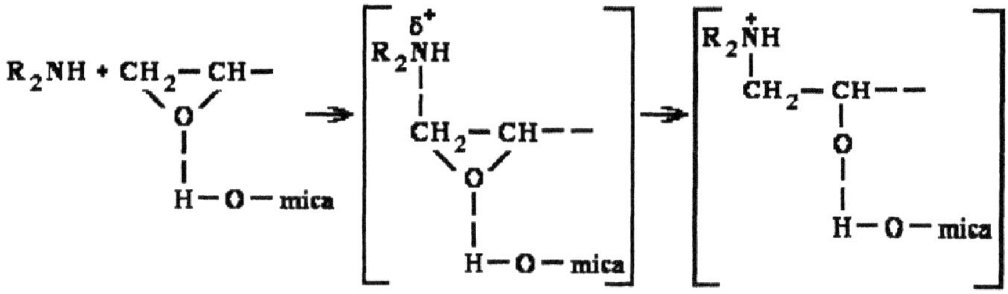

Here the reaction activation energy decreases from a value of 20 kcal/mol for a unfilled oligomer to 13.2 kcal/mol for an oligomer filled with 100 phr (parts by weight per hundred parts of the resin). Similar effect is observed during the polymerization of epoxy oligomers in the presence of a solid acid of a silica type [48].

The filler acid-base properties are effectively used when employing an arbitrary new method of olefine polymerization consisting in a so-called polymerization filling [49]. Here the acid- or the base-in-character sites serve for grafting of the complex catalysts of olefine polymerization on the filler surface, where kaolin, silica-gel, calcium and manganese carbonate and other compounds are used as fillers. Here olefine polymerization takes place on the activated filler surface. The method allows for the variation of the filler content in wide ranges - up to a very large filling percentage. It enables one to effectively regulate the molecular weight of the forming polymer and to obtain a structure homogeneity of the composites without using special dispersing admixtures.

There are few data in literature for the polymerization and polycondensation of acid-in-character monomers and oligomers in the presence of fillers - solid bases. For instance, the study [50] illustrates how calcium and magnesium oxide used as crosslinking additives significantly affect a composition of vinylchloride and acrylic acid. It is supposed that such copolymer grafting runs via the acid hydroxyl group of the acrylic bonds. This occurs thanks to the basic surface sites of calcium and magnesium oxides. The same effect is observed when using these oxides as densing additives for compositions that are based on polyester oligomers [51].

In that case, the basic oxides react with the polyester carboxyl groups according to the following mechanism [52]:

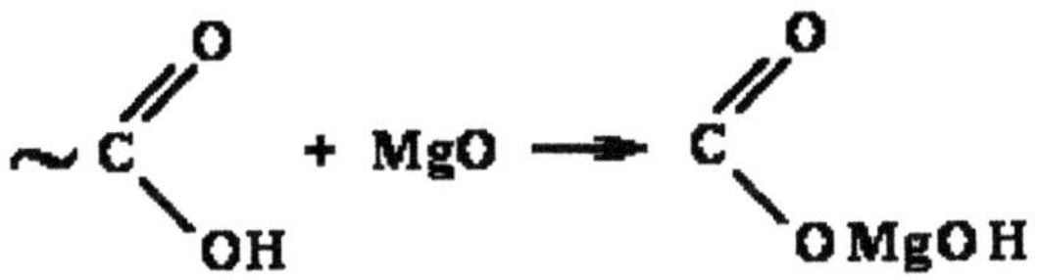

It is interesting to note that polymer basic-in-character groups may participate in the reaction at a later stage. These are the carbonyl groups while the metal plays the role of an electron acceptor:

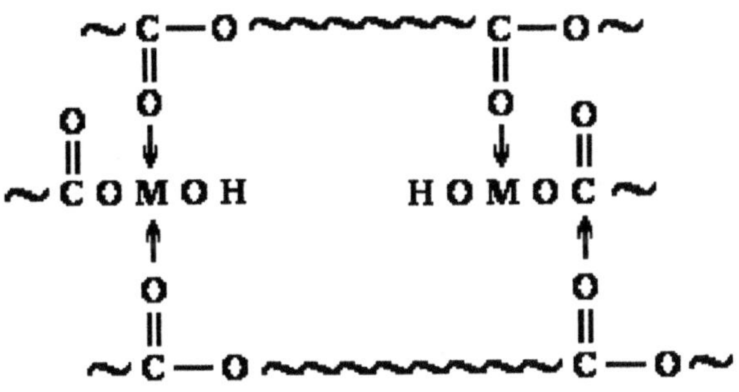

The results cited in [53] show that the improvement of the properties of polymer composites can be expected only in cases when a basic-in-character polymer is filled with acidic filler and vice versa. These ideas are plausibly formulated and proved in the studies of Fowkes [53], when studying adsorption of a polymer of basic-in-character (polymethylmethacrylate) and adsorption of a polymer of an acidic character [chlorinated poly(vinylchloride)]. Adsorption takes place on an acidic filler (silicon dioxide) and on a basic one (calcium carbonate), respectively, from acidic, neutral and basic solvents. It is shown that polymer adsorption on the fillers is determined by the acidity or basicity of polymers, solvents and filler surface sites. In addition, it is confirmed that the acid-base interaction is decisive for the establishment of a strong adhesion bond between filler and polymer. This fact should be considered when tackling polymer composite formation. [52, 53] give practical recommendations for the choice of inorganic pigments to be introduced into the polymer matrix when optimizing polymer coating fabrication and its mechanical properties. As noted in the studies, basic pigment should be introduced into acidic polymers and vice versa. Note another important point - when the active surface sites of the filler are blocked by introducing silanes, the mechanical characteristics of basic polymer coatings worsen, as compared to those of coatings where a "native" filler is introduced. Based on the obtained results, Fowkes concludes that the following principle should be accounted for when combining polymer and filler:

a) a basic-in-character filler should be introduced into an acidic polymer - for instance, polyvinylbutyral and chlorinated polyvinylchloride should be filled with glass that is processed with aminosilanes;
b) a basic polymer should be filled with an acidic filler - for instance, such a filler is glass processed with hydrochloric acid.

It is worth noting the important role of the acid-base interactions in the formation of polymer composites in the presence of ultrafine fillers. Those are the so-called hybrid organic-inorganic nanocomposites [54]. The ultrafine fillers improve the composite properties to a significantly larger extent in comparison with the more coarse fillers. Modified and unmodified layered silicates being solid acids are often used as nanofillers [55]. Two effects can be established caused by the interface interaction between basic monomers (acrylates, amides, etc.) and the acidic sites of the layered silicates. First, the process of monomer polymerization accelerates. Secondly, the silicate layer delamination into particles with nanodimensions becomes easier [56]. Although a large number of researches has been made on this topic, the specificity of the formation and rheological behaviour still needs for more elucidation [57-61].

REFERENCES

1. Kurokawa N., Tabata M., Sohma J. Polymerization of ethylene and propylene initiated by milled alumina powder. J. Appl. Polym. Sci. 1980; **25**: 1209-16.
2. Vuillume G., Spitz R., Revillon A., Guyot A. Supported chromium oxide catalysts for olefins polymerization. X. Propylene polymerization-activity measurements at low pressure. J. Macromol. Sci. 1974; **A8**: 1117-28.
3. Spitz R., Revillon A., Guyot A. Supported chromium oxide catalysts for olefin polymerization. XI. Comparison between ethylene and propylene polymerization. J. Macromol. Sci. 1974; **A8**: 1129-36.
4. Friedlander H.Z. Organized polymerization. I. Olefins on a clay surface. J. Polym. Sci. 1964; **C4**: 1291-301.

5. Take J., Ikeda M., Yoneda Y. Nature of catalytical active sites over solid acids. III. Relationships between acidic properties of silica-alumina and its catalytic activities for olefin polymerization. Bull. Chem. Soc. Jap. 1977; **50**: 49-51.
6. Fukano K., Kageyama E. Radiation-induced polymerization of styrene adsorbed on several inorganic substances. J. Polym. Sci.: Polym. Chem. Ed. 1975; **13**: 1309-24.
7. Fukano K., Kageyama E. Study of radiation-induced polymerization of vinyl monomers adsorbed on inorganic substances. VIII. Polymerization of styrene and methyl methacrylate adsorbed on aerosil. J. Polym. Sci.: Polym. Chem. Ed. 1976; **14**: 2183-92.
8. Arai M., Arai K., Saito S. Soapless emulsion polymerization of methyl methacrilate in water in the presence of calcium sulfite. J. Polym. Sci.: Polym. Chem. Ed. 1982; **20**: 1021-29.
9. Stav E., Bentur A., Kohn B.H. Polymeriztion of acrylamide in the presence of calcium sulfate dihydrate and calcium sulfate hemihydrate. J. Appl. Polym. Sci. 1992; **45**: 2079-89.
10. Stav E., Bentur A., Kohn B.H. Polymeriztion of acrylamide in the presence of calcium sulfate hemihydrate: compressive strength and microstructure of the composite. Adv. Chem. Res. 1991; **4**: 57-60.
11. Theng B.K., Walker G.F. Interactions of clay minerals with organic monomers. Isr. J. Chem. 1970; **8**: 417-24.
12. Solomon D.H., Swift J.D., Murphy A.J. The acidity of clay minerals in polymerization and related reactions. J. Macromol. Sci. 1971; **A5**: 587-601.
13. Polushkin V.A., Diner V.A., Lokutsievskii V.A., Khan I.G. Study of products of styrene polymerization initiated by kaolinite. Vysokomol. Soedin. 1984; **B26**: 420-25 (In Russian).
14. Litvinenko M.L., Diner V.A., Polushkin V.A. Copolymerization of monomers adsorbed on kaolin. Vysokomol. Soedin. 1986; **B28**: 16-20 (In Russian).
15. Pavlov S.A., Bruk M.A. Radiation-induced polymerization of monomers on the solid surface in systems with strong adsorption bonding. Vysokomol. Soedin. 1987; **A29**: 118-24 (In Russian).
16. Natta G., Mazzanti G., Pregaglia G., Binaghi M. Crystalline polymers of dialkyl ketenes. Macromol. Chem. 1961; **44-46**: 537-49.
17. Natta G., Mazzanti G., Pregaglia G., Binaghi M., Cambini M. Polymers of dimethylketene having essentially a polyacetal structure. Macromol. Chem. 1962; **51**: 148-53.
18. Cheshkov V., Natova M. Homogeneous-heterogeneous catalysis of the polycondensation process of amino resins. Proceedings of VI International Symposium on Relations between Homogeneous and Heterogeneous Catalysis; 1989 September 25 - 29; Piza, 1989.
19. Lipatova T.E., Sheinina L.S. On the effect of aerosil on the kinetics of formation of linear polyurethanes. Vysokomol. Soedin. 1976; **B18**: 44-7 (In Russian).
20. Lipatova T.E., Sheinina L.S., Vladimirova L.Yu., Maslak Yu.V. Effect of aerosil on structure and properties of filled network polyurethanes. Vysokomol. Soedin. 1987; **A29**: 747-52 (In Russian).
21. Faure M., Fraissard J., Imelik B. Etude des propertiétés superficielles de l'oxyde de magnesium. II. Polymérisation de l'oxide d'éthylene. Bull. Soc. Chim. France 1970; **8/9**: 2828-33.
22. Mund S.L., Bruk M.A., Abkin A.D. Some characteristics of the radiation-initiated homo- and copolymerization of vinyl acetate and acrylonitrile in an adsorption layer on an aerosil surface. Vysokomol. Soedin. 1976; **A18**: 2631-38 (In Russian).
23. Morozova E.M., Eliseeva V.I., Travnikova L.A. Effect of monomer-active filler system structurization on thin layer polymerization. J. Polym. Sci.: Polym. Symp. 1973; 73-7.
24. Yamaguchi T., Tanaka H., Mustafa A.B., Ono T., Ito H., Itabashi O., Endo M., Ohuchi M., Saito L. Polymerization of MMA in the presence of various metal oxides, sands and slags. Chem. Ind. 1974; **15**: 619-22.

25. Shaboldin V.I., Demishev V.N., Ionov Ju.A., Akatova S.P. Effect of mineral salts on the structure of liquid thiokol. Vysokomol. Soedin. 1982; **A24**: 1099-102 (In Russian).
26. Shaboldin V.I., Krashennikov A.I. Vulcanization of liquid carboxyl group-containing butadiene rubbers with transition metal salts. Vysokomol. Soedin. 1984; **B26**: 17-20 (In Russian).
27. Kruba L.E., Amfiteatrova T.A., Kozlov L.V. Effect of pigments on the flow and curing of thermosetting acrylic compositions. Lakokras. Mater. i ikh Primen. 1984; 14-16 (In Russian).
28. Kuznetsov G.K., Chirkova E.A. Effect of silicate fillers on the curing of polyester resins. Plast. Massy 1981; 20-22 (In Russian).
29. Tarasevich Yu.I., Rak V.C., Telichkun V.P. Study of the adsorption of amino acids on montmorillonite. Kolloidn. Zh. 1977; **39**: 1190-94 (In Russian).
30. Kozlov L.V., Alekseev S.M. Effect of the titanium dioxide surface on the polymerization of methyl methacrylate. Lakokras. Mater. i ikh Primen. 1974; 12-14 (In Russian).
31. Moustafa A.B., Radwan F.M., Aboul-wafa O., Emam A. Kinetics of the heterogeneous polymerization of methyl methacrylate catalized by unmodified and modified surface of the white sand of Sinai. Surf. Coat. Int. 1991; **74**: 107-10.
32. Moustafa A.B., Badran A.S. Effect of cations and anions of some metal salts in the aqueous polymerization of methyl methacrylate. Angew. Makromol. Chem. 1982; **103**: 153-58.
33. Talapatra S., Saha S.K., Chakravarti S. K., Guhaniyogi S.C. Aqueous polymerization of methyl methacrylate on hydrogen bentonite: effect of alcohols. Polym. Bull. 1983; **10**: 21-27.
34. Diab M.A., El-Sonbati A.Z., Hilali A.S., Killa H.M., Ghoneim M.M. Polymer complexes - X. Polymerization of methyl methacrylate in the presence of some transition metal chlorides. Eur. Polym. J. 1990; **26**: 1-3.
35. Spasskaya R.L., Zil'berman E.N. Polymerization of caprolactam in the presence of lead acetate. Vysokomol. Soedin. 1985; **B27**: 247-50 (In Russian).
36. Gorbunova Ye.V., Deev Yu.S., Ryabov Ye.A. Autopolymerization of capro- and dodecalactam. Vysokomol. Soedin. 1980; **A22**: 2457-63 (In Russian).
37. Gorbunova Ye.V., Kazaryan L.G., Tsvetkov V.D., Azriel A.Ye., Deev Yu.S. Kinetic features of polymerization of capro- and dodecalactam in the presence of copper, vanadium, manganese, molibdenum and cobalt oxides. Vysokomol. Soedin. 1988; **A30**: 342-47 (In Russian).
38. Gorbunova Ye.V., Shifrina R.R., Deev Yu.S., Ryabov Ye.A Study of features of polymerization of lactams initiated by transitional metal oxides. Vysokomol. Soedin. 1983; **A25**: 2605-10 (In Russian).
39. Gorbunova E.V., Deev Yu.S., Ryabov E.A. Mechanism of the polymerization of lactams in the presence of transition metal oxides. Plast. Massy 1980; **4**: 17-9 (In Russian).
40. Pilyugin B.G., Gorbunova E.V., Deev Yu.S. Heterogeneous polymerization of caprolactam and preparation of filled poliamide. Plast. Massy 1989; **4**: 14-7 (In Russian).
41. Cheshkov V.M., Natova M.M., Ashurov N.R., Usmanova M.M., Veksel'man A.M. Polymerization filling of polycaprolactam. Eur. Polym. J. 1991; **27**: 205-8.
42. Natova M.M., Cheshkov V.M., Ashurov N.R., Usmanova M.M., Veksel'man A.M. Polymerization-filled polycaprolactam. Eur. Polym. J. 1993; **29**: 653-55.
43. Arimoto H., Ishibashi M., Hirai M., Chatani Y. Crystal structure of the γ-form of nylon-6. J. Polym. Sci. 1965; **A3**: 317-26.
44. Vol'f L.A., Haikin B.S. *Caprolactam Polymerization* (*Polimerizatsija Kaprolaktama*). Leningrad: Leningr. Univ., 1982 (In Russian).
45. *Ring-Opening Polymerization*, Vol. 2,: K. J. Ivin and T. Saegusa, eds. London-N.Y.: Elsevier Appl. Sci. Publ. Ltd., 1984.

46. Tai K., Arai Y., Tagawa T. The simulation of hydrolytic polymerization of ε-caprolactam in various reactions. J. Appl. Polym. Sci. 1982; **27**: 731-46.
47. Rothe M., Reinish G., Jaeger W., Shopov I. Die kationische Caprolactam Polymerization. Makromol. Chem. 1962; **B54**: 183-88.
48. Bozveliev L.G. Peculiaritiy on the changes of sites, depending on disperse fillers contents. Thesisses DSc., Sofia: HICHE, 1989 (In Bulgarian).
49. Enikolopov N.S., Vol'fson S.A. Production and properties of filled thermoplastics. Plast. Massy 1988; **1**: 39-49 (In Russian).
50. Mozzhukhin V.B., Sadoka S.P., Fedoseev B.L., Rasmanik I.V. Effect of crosslinking additives on the adhesion strength of vinyl chloride-acrylic acid copolymer plastisol coatings. Vysokomol. Soedin. 1988; **B27**: 414-6 (In Russian).
51. Marmo M.J., Mostafa M.A., Jinnai H., Fowkes F.M., Manson J.A. Acid-base interaction in fiber-matrix systems. Int. Eng. Chem. Prod. R & D 1976; **15**: 206-11.
52. Mihaylova Z.V., Pugachevskaya N.F., Kovriga V.V. "Filled composition on the base of nonsaturated polyesters." In *Chemistry and Technology of High Molecular Compounds*, A. A. Askadsky, ed. Vol. 14, Moscow: VINITI, 1981 (In Russian).
53. Fowkes F.M. Role of acid-base interfacial bonding in adhesion. J. Adhes. Sci. and Technol. 1987; **1**: 7-27.
54. Komarneni S.S. Nanocomposites. Mater. Chem. 1992; **2**: 1219-30.
55. Wang M., PinnavaiaT. Clay-polymer nanocomposites formed from acidic derivatives of montmorilonite and an epoxy resin. Chem. Mater. 1994; **6**: 468-74.
56. Messersmith P., Giannelis E. Synthesis and characterization of layered silicate-epoxy nanocomposites. Chem. Mater. 1994; **6**: 1719-25.
57. Schmidt H. New type of non-crystalline solids between inorganic and organic materials. J. Non-Cryst. Solids 1985; **73**: 681-91.
58. Novak B. M. Hybrid nanocomposite materials - between inorganic glasses and organic polymers. Adv. Mater. 1993; **5**: 422-33.
59. Wang Z., Lan T., Pinnavaia T. Hybrid organic-inorganic nanocomposites formed from an epoxy polymer and a layered silicic acid. Chem. Mater. 1996; **8**: 2200-04.
60. Kotsilkova R., Natova M., Ruseva S. Sinthesing and structural characteristics of the epoxy/layered silicate hybrid nanocomposites. Phys.-Chem. Mechanics 1999; **25**: (in press) (In Bulgarian).
61. Kotsilkova R., Okamoto M., Taguchi H., Sato N., Kotaka T. "Shear and elongational flow of smectite/PMMA hybrids and glass fiber/PMMA composites." In *Progress and Trends in Rheology*, Igor Emri, ed. Darmstadt: Springer 1998; 398-99.

Chapter 5

PROPERTIES OF POLYMER COMPOSITES WITH ACIDIC OR BASIC FILLERS

As shown so far, the acid-base interactions significantly affect polymer formation when adding solid acids or bases. Besides, those interactions affect also polymer and composite characteristics. Regarding the character of the occurring acid-base interactions, their effect may be either positive or negative, i.e. they may either improve, or make the properties of the composite material worse. The effect of those interactions depends on the composition of each particular system – acidic or basic filler added to basic or acidic monomer, oligomer or polymer.

Blocking up of the polymer functional groups by their acid-base interaction with the filler surface may yield an increase of the polymer resistance against hydrolysis. The authors of study [1] establish that fact, analyzing composites based on polyurethane polymers.

When filling polymers with aerosil, one may observe significant improvement of the material mechanical characteristics – see Table 1. The following abbreviations are used: TDI – toluilene-2,4-diisocyanate, MDI – diphenyl-methane-4,4' with a second component – poly(oxypropylene glycol); φ - degree of filling, % by weight; σ_t – tensile strength, MPa; E_∞ - modulus of viscoelasticity, MPa.

The authors of [1] attribute the improvement of the polymer mechanical characteristics to an interaction of donor-acceptor type between the polymer basic-in-character groups and the hydroxyl groups of the fillers. Manson and Williams study the contribution of the acid-base interactions to the improvement of the mechanical properties of filled polymers [2, 3]. They analyze compositions based on copolymer of ethylene and vinylacetate and silicon dioxide (aerosil). Because of the vinylacetate groups, the copolymer has basic-in-character properties, while the properties of the aerosil are acidic due to the presence of silanol groups on its surface. The copolymer basicity was changed owing to the increase of the portion of basic-in-character groups (the concentration of vinylacetate varies within ranges 0–32 wt. %). Aerosil acidity was changed due to its surface treatment with silane, and such a treatment blocks most of the acid sites (more than 75 %). Table 2 shows the effect of the vinylacetate concentration in the polymer. It also shows how the treated or non-treated aerosil affects the mechanical characteristics of the polymer composite material, i.e. relations between the moduli E/E_0 and of the current and initial tensile strengths TS/TS_0 have changed.

Table 1. Mechanical properties of polyurethanes filled with aerosil

System with TDI			System with MDI		
φ	σ_t	E_∞	φ	σ_t	E_∞
0.0	1.5	4.0	0.0	2.9	3.8
1.0	1.7	2.0	1.0	3.4	4.0
6.5	4.0	1.7	6.5	8.9	8.2
10.5	5.8	4.6	10.5	14.8	9.1
13.0	5.6	4.1			

As seen, both the elastic moduli and the tensile strengths of the composite material increase with the intensification of the acid-base interactions. The improvement of the mechanical characteristics is a result of the increase of both the number of the filler surface acidic sites and the concentration of the polymer basic groups.

Table 2. Relative modulus (E/E_0) and tensile strength (TS/TS_0) of 8 vol. % silica-filled copolymers of ethylene and vinylacetate (VA)

Silica	Surface treatment	Content of VA, %	E/E_0	TS/TS_0
Aerosil 130	None	0	1.38	1.22
		10	2.15	1.44
		32	3.87	1.80
Aerosil R 972	Chlorosilane	0	1.44	1.11
		10	2.03	1.27
		32	2.26	1.49

When filling polyamide-6 with chlorides of lithium, calcium, iron, magnesium or manganese, one may observe a decrease in the polymer melting temperature, decrease in the crystallization rate and increase in the temperature of glass transition [4-12]. One may explain the change of the material characteristics in terms of the strong donor-acceptor interactions that take place between the acidic-in-character coordination unsaturated metal cations and the basic-in-character carbonyl groups of the polymer. Note that these metal salts are water-soluble. Hence, specific salt quantity dissolves in the water that the hydrophilic polyamide contains. The water solution of the salt diffuses in the polymer and interacts with the polymer basic-in-character groups. Thus, the number of interchain hydrogen bonds decreases. Such type of reaction yields a decrease of polyamide degree of crystallinity and deterioration of the polyamide physico-mechanical properties [8]. Such a phenomenon is most typical when moisture increases.

However, the behaviour of a system that consists of polyamide-6 and zirconium tetrafluoride is different [12]. The system contains a solid acid that is entirely water-insoluble, and a number of complexes form from it. For instance, there are stable complexes between the polymer amide groups and ZrF_4, and charged complexes of chain segments being part of the amorphous polymer area. The authors assume that a specific molecular reinforcement of the polymer takes place. It leads to an improvement of the polymer mechanical properties – see Table 3.

Regarding the properties of a unfilled polymer, Table 3 illustrates the improvement of mechanical characteristics of a polymer filled with solid acid. The properties that improve include the polymer tensile and flexural strength, the corresponding moduli and the ultimate elongation. Note that the values of the mechanical parameters increase as the concentration of ZrF_4 increases.

Table 3. Mechanical properties of polyamide-6 filled with ZrF_4

Content of ZrF_4, %	Tensile strength, 10^{-4} Pa	Modulus of elasticity, 10^{-4} Pa	Ultimate elongation, %	Flexural strength, 10^{-4} Pa	Modulus of elasticity, 10^{-4} Pa
0	6.96	27.30	6.40	10.76	34.49
5.00	9.45	34.54	7.10	11.38	39.03

As already discussed in Chapter 4, the introduction of solid acids into ε-caprolactam enables one to produce polycaprolactam via polymerization caused by the catalytic performance of the surface acid sites [13]. Thus, the introduced inorganic substances act both as catalysts of the polymerization of a basic-in-character ε-caprolactam and fillers of the resulting polycaprolactam. Heterogeneous polymerization reaction that runs in the presence of a filler with catalytic action is referred to as ***polymerization filling***, while the resulting product is a ***polymerization-filled polymer***.

We now discuss some intrinsic characteristics and structural peculiarities of a polymerization-filled polycaprolactam considering the experimental evidence gained in our laboratory. The polymer is prepared by using γ-alumina and calcium phosphate as solid acids. Table 4 lists the characteristics of the polymerization-filled polycaprolactam, and, for comparison, gives characteristics of various types of polycaprolactam, produced by conventional methods for homogeneous polymerization [4–10, 14–21]. The fillers used are distinguished by the strength of the filler acid sites, ranging from $pK_a = +2.0$ to $pK_a = +4.8$. Note that despite the filler high percentage, the polymerization-filled polycaprolactams are not inferior but even superior to known types of polycaprolactams, considering molecular weight, degree of crystallinity and dynamic modulus of elasticity.

Data in the last column are valid for unfilled polycaprolactams. The introduction of fillers, following the conventional procedure, often causes deterioration of the polymer properties. For instance, [19] reports that the addition of 15 % graphite to conventional polycaprolactam yields decrease in the tensile strength from 60 MPa to 28 MPa. Reimschuessel has established a decrease in the melting temperature from 222 K to 204 K when 15 % of ZrF_4 is added, while the crystallization temperature drops from 440 K to 423 K [9, 12]. When introducing 12,5 % of $FeCl_3$ in the polycaprolactam, the degree of crystallinity drops from 31 % to 16 % [20]. The comparison of those data with our experimental evidence confirms the efficiency of the method of polymerization filling.

It is impossible to correlate all the characteristics of the composite materials to the strength of the acid sites located on the inorganic surfaces. Data in Table 4 show that catalysts with stronger acid sites produce caprolactams with higher glass transition temperature, as compared to catalysts with weaker sites and lower molecular weight. Thus, a system with γ-alumina containing acid sites of $pK_a = +2.0$ up to $+3.0$ yields a polymer with molecular weight of 30000 - 36000 (depending on the filler amount). Here $T_g = 340$ K. However, the resulting polymer has molecular weight of 43000 and $T_g = 327$ K when calcium phosphate ($pK_a = +3.3$ up to $+4.8$) is added. For mixed catalysts, the polycaprolactams obtained have similar characteristics, except for the molecular weight that is larger than the molecular weight of a single filler. At the same time, melting and crystallization temperature, as well as the degree of crystallinity of all polymerization-filled polycaprolactam samples, are comparable.

The studied samples have a single peak of melting, regardless of the acidity of the filler-catalysts introduced. Peak location (in the range 492 – 494 K) is characteristic of the polycaprolactam crystal α-form that is thermodynamically most stable. Annealing (widely used in practice) at $T = 458$ K for 20 hours leads to a strong increase in both the values of the dynamic modulus of elasticity and the degree of crystallinity. The glass transition temperature also increases to a certain extent.

Fig. 1 illustrates some typical changes of the dynamic modulus of elasticity of polymerization-filled polycaprolactams (annealed and unannealed) as a function of temperature. This figure records the most significant changes of the modulus, taking place within the temperature range of glass transition.

Table 4. Characterization of polycaprolactams

	Heterogeneous polymerization (polymerization filling)				**Homogeneous polymerization**
Filler	**$Ca_3(PO_4)_2$, 30 wt. %**	**γ-Al_2O_3, 5 wt. %**	**γ-Al_2O_3, 10 wt. %**	**γ-Al_2O_3, 5 wt. % + +$Ca_3(PO_4)_2$, 20 wt. %**	
Properties					
Molecular weight, M_ηx10^{-3}	43	36	30	51	15-25
Melting temperature, T_m, K	492	494	494	492	491-498
Glass transition temperature, T_g, K	327	340	338	333	323-330
Crystallization temperature, T_c, K	454	457	457	456	420-440
Degree of crystallinity, %	39 64	31 40	34 40	31 37	22-33
Tensile strength, MPa	58	62	64	52	40-60
Dynamic modulus of elasticity, E_d.x10^{-2}, MPa	76 132	59 79	65 84	82 120	55-60

Note: * When there are two values, the second refers to annealed samples (458 K, 20 h).

Fig. 2 gives plots of the tangent of the angle of mechanical loss vs. temperature for annealed and unannealed samples of the composite material. The plots display molecular mobility within the polymer crystalline and amorphous area. One may also observe two main temperature transitions that are characteristic of polycaprolactams prepared by employing conventional technology. The first one refers to the molecular mobility of the amide groups at 205 – 225 K not bound trough hydrogen bonds. The second temperature transition (at 320 – 345 K) can be referred to the segmental mobility of the polymer amorphous phase. Hence, that transition gives the glass transition temperature.

Fig. 3 presents X-ray diffraction patterns typical of all samples of the composite material. Two peaks are present - at $2\theta = 20.2\,^0$ and $2\theta = 23.6\,^0$. Both they are characteristic of α-crystal modification of polycaprolactam (bulk samples). A third peak at $2\theta = 21.2\,^0$ is characteristic of γ^*-modification (surface layer) [8]. It also characterizes the stable γ-crystal form of the polycaprolactam. However, we should also expect a peak at $2\theta = 11\,^0$, but it is not present in our case. Note that the presence of a peak at $2\theta = 21.2\,^0$ confirms the assumption that an unstable γ^*-modification exists.

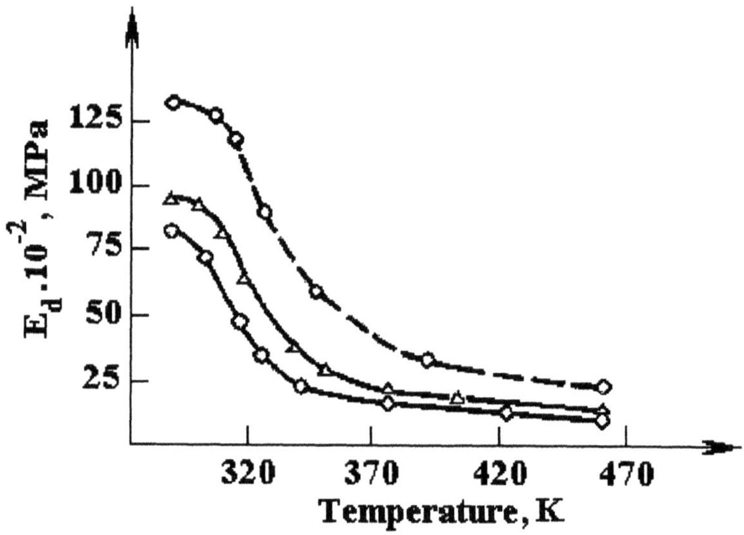

Figure 1. Dynamic elastic modulus *vs.* temperature of the polymerization filled polycaprolactam samples: —Δ— - 10 wt. % γ-Al_2O_3 ; —O— - 30 wt. % $Ca_3(PO_4)_2$; --- O --- - 30 wt. % $Ca_3(PO_4)_2$, annealed

In fact, the peak at $2\theta = 21.2\,^0$ disappears under sample annealing, and two new peaks occur, at $2\theta = 20.2\,^0$ and $2\theta = 23.6\,^0$. This means that a transition of the γ^*-modification to α-crystal form takes place. The presence of various crystal modifications can be attributed to the varying conditions of crystallization, being characteristic of the preparation of different types of composite material. The so-called ***ad-wall*** effect is thought to happen in between. Another note-worthy phenomenon is the occurrence of a sharp peak at $2\theta = 25\,^0$ in all diffraction patterns of the polymerization-filled polycaprolactams. It corresponds to an interplanar distance of 3.55 Å. Note that there is no such peak in the diffraction patterns of the applied filler-catalysts. The same is true for the patterns of the known crystal forms of a polycaprolactam. Hence, the X-ray peak occurrence can be associated with a definite polymer ordering (i.e. structurizing) at the polymer/filler particle interface. The interplanar distance is a unique value for any crystal body and it appears as a very narrow diffraction maximum on the x-ray diagrams. In our case this fact can be thought to occur due to the quite regular orientation of polymer molecules towards the surface of the active filler. We may assume that this ordering resembles crystallization of an epitaxial type. The acid sites of the filler surface determine such a type of ordering. Under sample annealing, however, intensity of the diffraction maximum decreases. One may explain this phenomenon with a partial disorientation of polymer molecules on the interface.

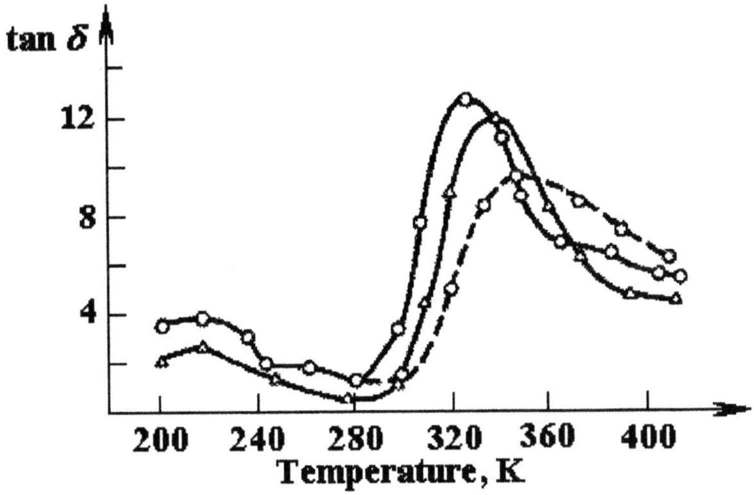

Figure 2. Tangent of the angle of mechanical loss *vs.* temperature dependence for polymerization filled polycaprolactam samples: —Δ—- 10 wt. % γ-Al_2O_3; —O—- 30 wt. % $Ca_3(PO_4)_2$; --- O --- - 30 wt. % $Ca_3(PO_4)_2$, annealed

It was also found that the physico-mechanical characteristics of polymerization-filled polycaprolactams can be further improved through the introduction of fiber fillers (basalt fibers). Note that basalt fibers are inert, regarding polymerization. Table 5 illustrates the improvement of polymer physico-mechanical properties.

The submitted data show that the increase of the concentration of basalt fibers yields significant increases in the tensile strength, dynamic modulus of elasticity and degree of crystallinity. The sample melting temperature shifts to higher values. Note that those fibers, regardless of their concentration within the considered limits, do not inhibit catalytic effect of the basic filler (being polymerization catalyst).

Rao, Forssberg and Forsling observe an increase of the elasticity modulus by 66 % during tension of polymethylmetacrylate filled with wollastonite. They compare their data to the value of the modulus of elasticity of an unfilled polymer [21]. Obviously, this effect can be related to the high adhesion strength between the polymer and filler. Note that the acid sites (the hydroxyl groups), located on the wollastonite surface, interact with the PMMA basic-in-character groups (carbonyl ones). This determines the high polymer-filler adhesion strength. The effect of wollastonite modification by processing the compound with stearic acid is even stronger. Thus, the tensile modulus may increase by 78 % and the tensile strength – by 17 %.The authors explain the phenomenon with the additional interaction between the PMMA and the hydrocarbons of the stearic acid. At the same time, an effective linkage of the stearic acid with the wollastonite takes place. It can be assumed that this phenomenon is due to the interaction of the stearic acid with the basic-in-character sites located on the filler surface (together with the acid sites) [21]. Note that polymer compatibility with modified filler is better than that with non-modified filler. This explains the improvement of the polymer characteristics.

[22] considers a system consisting of vinylchloride/acrylic acid copolymer (VC – AA) and magnesium, calcium and lead oxide. It was found that the system admits acid-base interaction between the acidic polymer and the basic-in-character filler. The interaction results in polymer cross-linking and increasing of the polymer stiffness (Fig. 4).

The final result is significant increase of the adhesion strength of polymer to the aluminium foil (Fig. 5).

Table 5. Polycaprolactam, polymerization-filled with basalt fibers (the catalyst is 10 % wt. γ-Al_2O_3)

Properties	**Content of basalt fibers**			
	0	**0.3 wt. %**	**1 wt. %**	**5 wt. %**
Tensile strength, MPa	64	75	82	86
Dynamic modulus of elasticity, E_dx10^{-2}, MPa	65	70	74	78
Degree of crystallinity, %	34	37	40	42
Melting temperature, K	494	497	497	497

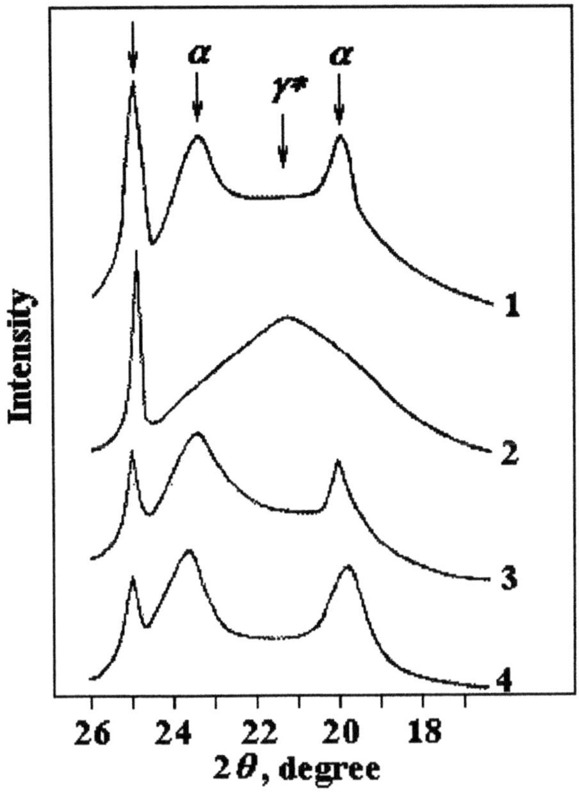

Figure 3. Typical X-ray diffraction pattern of polymerization filled polycaprolactam [30 wt. % $Ca_3(PO_4)_2$]: (*1*, *3*) - interior parts of samples before and after annealing, respectively; (*2*, *4*) - surface layers before and after annealing, respectively

In Chapter 4 we showed that aprotic solid acids, such as the sulphates of alkaline-earth metals, are strong catalysts of the polycondensation of amino-aldehyde oligomers. Regarding parameters of an unfilled polymer or those of a polymer filled with inactive fillers, the structure and properties of the forming polymers significantly change here. As is known, the basic disadvantage of amino-aldehyde polymers, as compared to other polymers, consists in their extremely low durability. This is due to the very strong polymer destruction in time and is most typical for amino-formaldehyde polymers, cast under atmospheric pressure and at room temperature [23-31]. The low material durability significantly restricts the practical application of those polymers, regardless of their valuable properties – strong adhesion to both organic and inorganic templates, high surface hardness, light- and biological resistance. On the other hand, amino-formaldehyde polymers have some advantages.

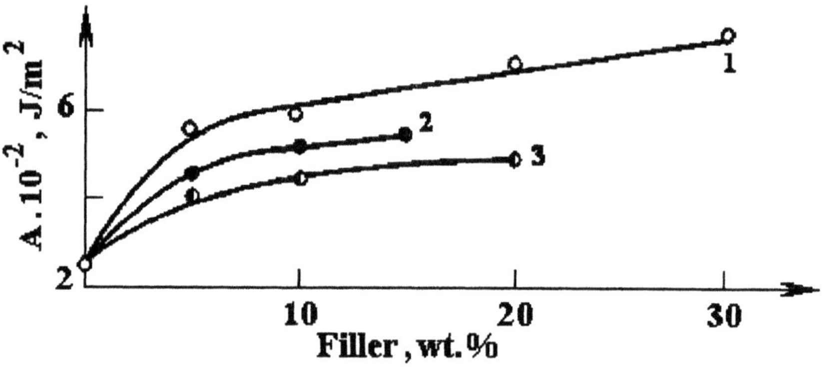

Figure 4. Effect of CaO (*1*), PbO (*2*) and MgO (*3*) on the stiffness of films based on copolymer VC – AA

- accessibility and low cost of the raw materials and chemical precursors needed for their preparation - mainly carbamide, melamine and formaldehyde;
- their oligomers mix well with water;
- easy oligomer-polymer transition not only at high but at room temperature.

Hence, one should not neglect the practical importance of those composite materials.

Due to these disadvantages the amino-formaldehyde polymers are scarcely used in pure form and are most often combined with other polymers. Pure amino-formaldehyde polymers are often used as clays in wood-processing industry due to the strong polymer/wood particles adhesion [23-27]. A number of factors determine the intensive destruction and the tendency to cracking of amino-formaldehyde polymers. These are as follows:

- the presence in their structure of functional groups that can easily hydrolyze;
- very high polydispersity of the supermolecular structure;
- high density of cross-linking and low elasticity of the macromolecules;
- large internal stresses occurring as a result of the significant volume contraction during the oligomer-polymer transition.

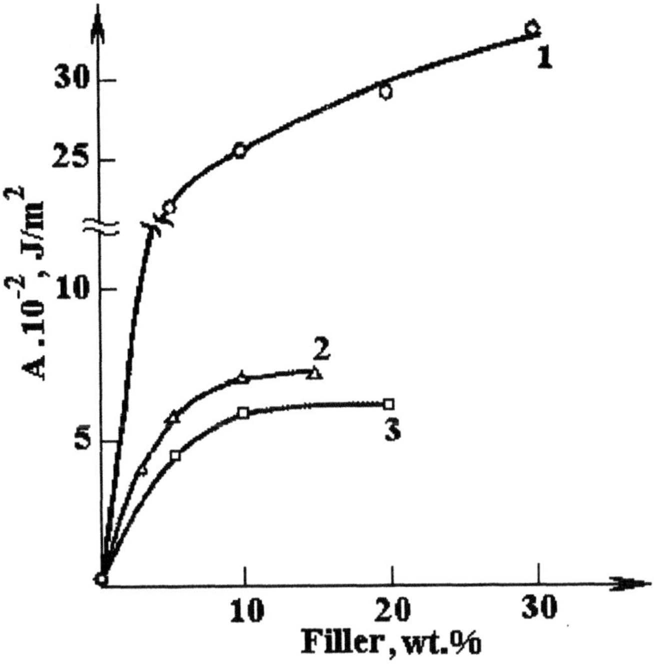

Figure 5. Effect of CaO (*1*), PbO (*2*) and MgO (*3*) on the strength of exfoliation of coatings based on VC – AA

Note that these factors are interrelated in a complex manner. Hence, it is impossible to simply separate a cause from an effect. The effects of the various factors on the amino-formaldehyde polymer durability vary significantly and depend on the conditions of oligomer-polymer transformation. Additional factors also affect polymer durability: the nature and concentration of the catalysts, the nature of the initial oligomers, and the parameters of polymer synthesis.

Table 6 presents data obtained throughout our study. We considered the change of strength behaviour of amino-formaldehyde polymers (non-filled polymers and polymers filled with solid acids). Our tests were performed under atmospheric pressure and at room temperature. To give a better demonstration of the effect of filling, we chose amino-aldehyde polymers that have low or zero durability, if unfilled. We prepared them from non-modified water-soluble urea and melamine-formaldehyde oligomers, without applying any pressure and at room temperature. The ratio oligomer/filler is 1:0.6 by volume. Data show that non-filled polymers destruct spontaneously soon after their preparation.

Solid acids, such as silicon and aluminium oxides and calcium phosphate, have only short-term effect on the polymer strength.

Table 6. Change of the bending strength in time for polymer materials based on urea-formaldehyde (UFO) and melamine-formaldehyde (MFO) oligomers

Composition	σ_B, MPa, for τ, days						
	8	**20**	**30**	**90**	**270**	**360**	**5400**
UFO	**D***-	-	-	-	-	-	-
UFO + SiO_2 (quartz)	24	25	18	6	3	**D**	-
UFO + $Al_2O_3.2SiO_2.2H_2O$ (kaolin)	25	**D**	-	-	-	-	-
UFO + Al_2O_3	10	-	3	**D**	-	-	-
UFO + TiO_2	12	7	5	2	**D**	-	-
UFO + $Ca_3(PO_4)_2$	18	15	14	10	5	3	**D**
UFO + $BaSO_4$	17	18	18	19	18	20	20
UFO + $SrSO_4$	27	28	-	27	28	28	28
UFO + $CaSO_4$	20	23	-	24	23	23	23
UFO + $CaSO_4.2H_2O$	21	23	23	23	23	23	23
MFO	-	**D**	-	-	-	-	-
MFO + SiO_2 (quartz)	-	-	5	3	**D**	-	-
MFO + $CaSO_4.2H_2O$	-	-	14	14	15	-	16

Note: * **D** - spontaneous destruction.

Based on the submitted data, we conclude that these fillers only restrict the crack *propagation* in the material, creating stearic obstacles, but do not eliminate the cause of crack formation and polymer destruction. Solid acids (sulphates of alkali metals) used as fillers, change the polymer structure; the strength characteristics of the composite materials do not vary in time, and the tendency of amino-aldehyde polymers to crack and destruct is completely eliminated. This is due to the significant structural changes that take place in filled polymers, in contrast to non-filled polymers and to polymers with inactive fillers. We considered amino-aldehyde polymers, filled with sulphates of alkali-earth metals [32-35]. We found that the polymer retains a linear structure that is typical for the oligomer and has fewer cross-linking bonds. The increase of the relative intensity of the absorption band in the infrared spectrum (at 1250 cm^{-1}) confirms this phenomenon. Note that this absorption band is characteristic for the scissoring oscillations of the imide groups of the linear grouping (Fig. 6). The intensity increase is a result of the reaction catalysis that decreases the probability of structural cross-linking at the nitrogen atom. The reaction catalysis maintains secondary groups constant, and to a great extent the ternary ones also.

The electron-microscope photographs show a decrease in the size of the supermolecular formations and a decrease in their polydispersity; they also suggest a certain fibrillation of the polymer structure. Thus, increase in the deformation barrier against crack formation or increase in the critical deformation takes place, and the polymer deforms without cracking until it attains the critical strain. On the other hand, cyclic groupings of uronic type, i.e. weak points of the polymer structure, disappear. The lack of adsorption bands at 835 cm^{-1} proves this structure specificity. Thirdly, the number of labile methylol groups in material decreases; this fact is proved by the transition of the absorption bands at 1020 cm^{-1} into a spectral shoulder.

Thus the structure is changing by the formation of crack-resistant amino-aldehyde polymers. Those changes can be associated with the adsorption interaction between oligomer molecules and the surface acid sites of the filler. The entire blockage of the acid sites arises from the introduction of basic-in-character surfactants. This phenomenon yields:

a) decrease of the polycondensation rate; (ii) increase of the overall activation energy;
b) crack-resistance of the obtained polymer. Hence, it became clear that a specific transition of the catalytic filler has been transformed into an inactive one.

We have suggested the possibility of using sulphates of alkaline-earth metals as fillers, thus getting crack-resistant amino-aldehyde polymers. Fig. 7 gives results on performance characteristics of a composite material obtained from urea-formaldehyde oligomer at room temperature and under atmospheric pressure. It also gives the possible ranges of the material properties. Note that the filler-catalyst of polycondensation is calcium sulphate dihydrate. Introduction of other inactive fillers may also vary material density and mechanical properties.

Consider three important for the practice urea-formaldehyde composites. The first one consists of urea-formaldehyde oligomer, calcium sulphate dihydrate and silicon dioxide (quartz). It has the following mechanical characteristics - density of 2000 kg/m^3, bending strength – 25 MPa, compression strength – 70 MPa and shock resistance – $2.6.10^4$ J/m^2. The stability coefficient was established in a corrosion-aggressive medium consisting of 5 % solution of NaOH and 10 % solution of H_2SO_4 for 360 days. The corresponding values of the stability coefficient, characterizing the material mechanical characteristics, are as follows: 0.62 and 0.85 for bending strength, 0.62 and 0.84 for compression strength and 0.88 and 0.96 for shock resistance. Thus, the composite can be used for the fabrication of corrosion-resistant products.

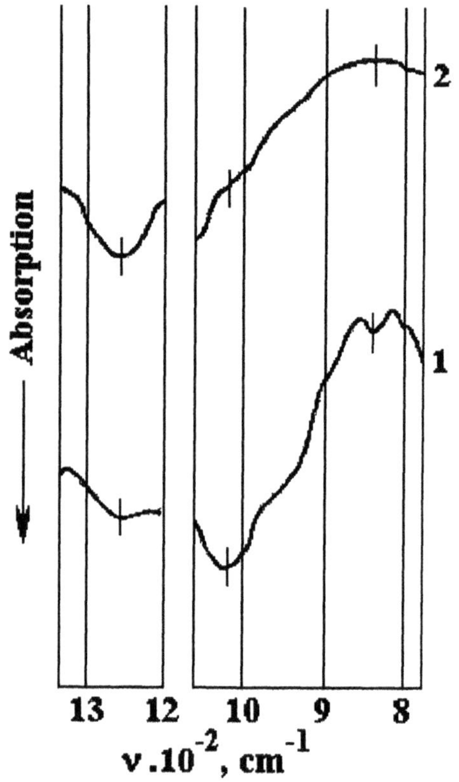

Figure 6. Infrared spectra of a urea-formaldehyde polymer obtained in the presence and in the absence of barium sulphate - curves (*1*) and (*2*), respectively

The second composition consists of urea-formaldehyde oligomer, calcium sulphate dihydrate and microspheres of fly-ashes from a thermal power station (TPS). Note that the walls of those TPS refuse consist of silicon, aluminium and ferric oxides, and they are hollow bodies filled with carbon and nitrogen dioxide. The composite material characteristics are as follows: density is in the range of 300 - 450 kg/m^3, compression strength is up to 6.5 MPa, bending strength is up to 3.5 MPa, modulus of elasticity is not larger than 300 MPa and the coefficient of thermal conductivity is in the range from 0.05 to 0.06 W/m.K; the composition could be used as an effective thermal-insulating material.

The third composition comprises urea-formaldehyde oligomer and barium sulphate as a filler-catalyst. The filling percentage is high, attained by using a surfactant. The composite material strongly absorbs γ-rays. Table 7 presents for comparison data on concrete intended for building of radiation protection defences.

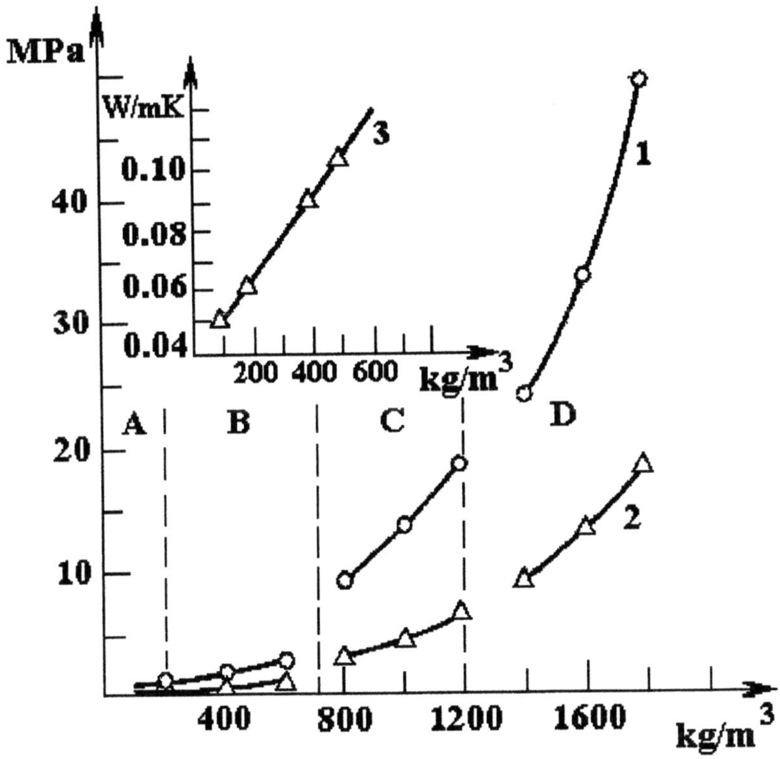

Figure 7. Ranges of the properties of a composition consisting of urea-formaldehyde polymer and calcium sulphate dihydrate: *1* – compression strength; *2* – bending strength; *3* – heat conductivity

Table 7. Weakening of γ-rays in a composite material based on amine-aldehyde polymer and barium sulphate

Parameter	Lead equivalent, mm			Linear coefficient of weakening, cm^{-1}		
	X-ray irradiation, 140 keV	Cs^{137}, 661 keV	Co^{60}, 1250 keV	X-ray irradiation, 140 keV	Cs^{137}, 661 keV	Co^{60}, 1250 keV
Composition	4.9	1.28	1.79	0.82	0.207	0.155
Concrete	1.9	1.08	1.58	0.33	0.169	0.132

The examples show that filler-catalysts can be used for the production of various composite materials. The fillers catalyze polycondensation of amino-aldehyde oligomers. Our results on wood-polymer composite materials are most illustrative [36-39]. As is known, those materials are widely applied in practice. One may produce them by using amino oligomers as binders. The materials, however, have some disadvantages, based on the basic disadvantage of the amino polymers, i.e. too low crack resistance of the polymer.

Consider formaldehyde separation that occurs during the exploitation of wood-polymer composites. It yields instability of the material mechanical characteristics in time, high water absorption and low water impermeability [28, 29].

Table 8. Properties of wood-polymer composite materials

Material / **Parameters**	$\mathbf{A_c}$	**A**	$\mathbf{B_c}$	**B**
Transversal tensile strength, MPa	1.7	1.7	0.5	0.7
Swelling in water, 24 h	4	2	6	1
Content of free formaldehyde, mg/100 g material	152	12	8	0
Weight loss during burning, %	20	5	20	3
Weight loss in a bio-aggressive medium, 6 months, %	44	1	44	3

Table 8 submits parameters that usually characterize the properties of wood-polymer composite materials. The material characteristics are obtained in the absence, and in the presence, of solid acid – calcium sulphate dihydrate. Material *A* is based on urea-formaldehyde oligomer where the content of free formaldehyde is high – up to 1 %. Such a material is prepared by using flat sheet pressing technology. Material *B* is based on urea-melamine-phenol-formadehyde oligomer with a low content of free formaldehyde (under 0.1 %). This material is prepared by applying extrusion pressing. The index *c* denotes materials prepared without introducing calcium sulphate dihydrate.

The table shows that the presence of active filler significantly improves the properties of the composite materials. It is important to note that, unlike the traditionally produced materials, the physico-mechanical characteristics of the new composites remain stable in time, even under varying temperature-moisture conditions. Note also that stable mechanical properties of wood-polymer materials can be obtained by using various active fillers – sulphates of alkali-earth metals and some industrial wastes (i. e. phosphogypsum). Hence, the used technology may have favorable ecological effect.

Consider transition of a filled epoxy oligomer into a polymer. One may violate the stoichiometric ratio between the oligomer and hardener functional groups, both participating in the polymer network formation [40]. The authors of the study prove that part of the acid sites, located on the surface of the glass microspheres, cannot participate in the reaction with the hardener; this follows from the interaction of the epoxy groups of the epoxy oligomer with the acid sites. Thus, the temperature of polymer glass transition decreases by 20 K as compared to the subsequent temperature of a non-filled polymer. Hence, a selective adsorption acid-base interaction takes place; it bonds the filler particles to the components of the subsequently formed polymer. The selective acid-base interaction may violate an important property of the composite material such as is its thermal resistance. One should take this effect into account when producing filled cross-linked polymers.

Table 9. Composition and properties of polyesters filled with $Al(OH)_3$

Parameters	**$Al(OH)_3$**		
	Untreated	**Treated with silane**	
Material composition			
Polyester resin based on isophthalic acid, weight parts	55	55	5560
Polyester resin containing bromine, weight particles	5	5	-
Contents of Br in polyester resins, %	2	2	0
$Al(OH)_3$, weight parts	175	175	200
Other components, weight parts	74.3	74.3	74
Material properties			
Viscosity, cPa	64000	33000	45000
Paste type	inhomogeneous	homogeneous	homogeneous
Stress at fracture, kg/m^2			
Tensile stress	546	597	509
Bending stress	1070	1390	1220

[41] submits interesting data on the effect of kaolin, talc, $Al(OH)_3$ and SiO_2, introduced as fillers in polyester composites. The study proves that the water impermeability and chemical resistance of cured composites increase. The improvement of the material properties is a result of the formation of strong linkages on the polymer–filler interface. The authors also present data on the decrease of the shrinkage of a curing polyester composite filled with kaolin.

The effect of shrinkage is significantly stronger in that case, even for 20 – 40 % filler content per 100 weight parts of the resin, as compared to 80 % polymer filling with $CaCO_3$ and glass spheres.

To regulate the properties of the polymer-filler interface, the filler surface can be treated. Silanes are among compounds that are most often used for filler surface treatment. For instance, the silane treatment of $Al(OH)_3$ provides increased fire resistance of the compound. Regarding $Al(OH)_3$ that has not been treated, one may introduce the compound in a polyester resin in a much larger proportion, without affecting the composite viscosity [41]. At the same time, the strength of the composite material significantly increases – see Table 9.

We should note here that the composites with silane treated $Al(OH)_3$ display good technological properties and performance characteristics after being processed.

Table 10. Treatment of the kaolin surface with silanes and effects on the properties of the composite material

Parameters	Untreated kaolin	Type of silane used for the treatment of kaolin		
		*A–174**	*A-163*	**Methylsilane**
Content of kaolin, weight parts per 100 weight parts of pure resin	80	100	100	100
Viscosity, cPa				
Initial	$3.2.10^6$	$0.29.10^6$	$0.21.10^6$	$0.16.10^6$
Final	$46.4.10^6$	$54.4.10^6$	$64.0.10^6$	$44.0.10^6$
Time for attaining the final viscosity, days	10	2	2	2
Bending strength, kg/cm^2				
Initial	682	865	703	820
After boiling for 8 hrs	555	745	810	746

Notes: * *A–174* – methacryl-oxypropyl-trimethyl siloxane
A–163 – methyl-trimethoxy silane

Silanes can be also effectively used for the improvement of the interphase interactions and properties of composite materials filled with kaolin. The introduction of kaolin in the material significantly increases the viscosity of the composite. Hence, it is found that when kaolin is introduced into polyester composites, the filling degree is more or less restricted. Processing of the kaolin surface with silane enables one not only to increase the filling degree, but also to improve the material corrosion resistance – see Table 10.

[41] presents a detailed analysis of how the surface modification affects the properties of alumina filled allylester composites. The authors perform a dynamic mechanical analysis of those compounds, treating them with four silane coupling agents of different concentrations. They determine the changes in properties by measuring the glass transition temperature, and the values of the moduli of elasticity of both the glassy and viscoelastic composites. They compare the thermomechanical behaviour of native allylester polymers with that of composites filled with unmodified or modified alumina. For optimum concentration of silane coupling agents in each system, the authors have found that the cross-linking reaction is most important. Moreover, maximum glass-transition temperature is one and the same for one and the same degree of effective surface coverage. The structure of each silane coupling agent significantly influences the glass-transition temperature, storage modulus and *tan δ*.

REFERENCES

1. Lipatova T.E., Sheinina L.S., Vladimirova L.Yu., Maslak Yu.V. Features of aerosil effect on structure and properties of filled network polyurethanes. Vysokomol. Soedin. 1987; **4**: 747-52 (In Russian).
2. Manson J.A., Marmo M., Williams J.T. Acid-base interactions and mechanical behaviour in filled polymers. Org. Coat. Plast. Chem. 1980; **42**: 190-94.
3. Manson J.A. Interfacial effects in composites. Pure and Appl. Chem. 1985; **57**: 1667-78.
4. Valenti B., Bianchi E., Greppi G., Tealdi A., Cifferi A. Bulk properties of synthetic polymer-inorganic salt systems. Melting behaviour of salted poly(caproamide). J. Phys. Chem. 1973; **77**: 388-95.
5. Valenti B., Bianchi E., Tealdi A., Russo S., Cifferi A. Bulk properties of synthetic polymer-inorganic salt systems. IV. Role of the polymer substrate. Macromolecules 1976; **9**: 117-22.
6. La Mantia F.P., Acierno D. Glass transition behavior of salted nylon 6. Colloid and Polymer Sci. 1981; **259**: 693-96.
7. Siegmann A., Baraam Z. Effect of metal halides on the glass transition temperature of nylon 6. Makromol. Chem. Rapid Commun. 1980; **1**: 113-17.
8. Cifferi A., Bianchi E., Marchese F., Tealdi A. Differential scanning calorimetry of poly(caprolactam)/inorganic salt systems. Makromol. Chem. 1971; **150**: 265-70 (In Russian).
9. Reimschuessel A.C., Kim Y.J. Stress-cracking of nylons induced by zink chloride solutions. Mater. Sci. 1978; **13**: 243-52.
10. Acierno D., Bianchi E., Cifferi A., de Cindio B., Migliaresi C., Nicolais L. Bulk properties of synthetic polymer-inorganic salt systems. III. Flow behavior and glass transition of salted polycaproamide. J. Polym. Sci., Polym. Symp. 1976; **54**: 259-69.
11. Votteler Ch., Hoffman V., Trafara G. Modifications of polyamide-6 and polyamide-4 intercalated with alkali halides. Makromol. Chem. 1984; **185**: 1953-77.
12. Reimschuessel H.K. Polymer-metal halide system. Colloid and Polym. Sci. 1982; **260**: 842-50.

13. Cheshkov V.M., Natova M.M., Ashurov N.R., Usmanova M.M., Veksel'man A.A. Characteristics of polymerization-filled polycaprolactam. J. Appl. Polym. Sci. 1997; **64**: 1255-58.
14. Fichera A., Malta V., Marega C., Zannetti R. Temperature dependence of the polymorphous phases of nylon 6. Makromol. Chem. 1988; **189**: 1561-67.
15. Gurato G., Fichera A., Grandi F.Z., Zannetti R., Canal P. Crystallinity and polymorphysm of 6-polyamide. Makromol. Chem. 1975; **175**: 953-75.
16. Kobljakov A.I., Barteneva A.G. Effect of relaxation transition on the strength of Kapron fibers. Vysokomol. Soedin. 1986; **A28**: 785-89 (In Russian).
17. Kojima Y., Usuki A., Kawasumi M., Okada A., Karauchi T., Kamigaito O. Synthesis of nylon-6-clay hybrid by montmorillonite intercalated with ε-caprolactam. J. Polym. Sci., Part A: Polym. Chem. 1993; **31**: 983-86.
18. Kojima Y., Usuki A., Kawasumi M., Okada A., Fukushina Y., Karauchi T., Kamigaito O. Mechanical properties of nylon-6-clay hybrid. J. Mater. Res. 1993; **8**: 1185-89.
19. Abdurazakov M., Gomza Yu.P., Petrenko S.D., Volkov V.N., Ashurov N.R., Shilov V.V., Lipatov Yu.S. Physicomechanical properties and structure of graphite-filled polycaproamide. Mekh. Kompoz. Mater. 1988; **2**: 257-63 (In Russian).
20. Chao L.C., Chang E.-P. Interaction of anhydrous ferric chloride with nylon 6. J. Appl. Polym. Sci. 1981; **26**: 603-10.
21. Rao K.H., Forssberg K.S.E., Forsling W. Interfacial interactions and mechanical properties of mineral filled polymer composites: wollastonite in PMMA polymer matrix. Coll. and Surf. 1998; **133**: 107-17.
22. Mozzhukhin V.B., Sadova S.P., Fedoseev R.L., Rasmanik I.V. Effect of additives on the adhesion strength of vynil chloride-acrylic acid copolymer plastisol coatings. Vysokomol. Soedin. 1985; **B27**: 414-16 (In Russian).
23. Dunker A.K., John W.E., Rammon R., Farmer B., Johns S.J. Slightly bizzare protein chemistry: urea-formaldehyde resin from a biochemical perspective. J. Adhesion 1986; **19**: 153-76.
24. Rammon R., John W.E., Magnusson J., Dunker A.K. The chemical structure of UF resins. J. Adhesion 1986; **19**: 115-35.
25. Mayer, Beat, *Urea-Formaldehyde Resins*. Massachusetts: Addison-Wesley Publ. Co., 1979.
26. Pizzy, Antonio, *Wood Adhesives: Chemistry and Technology*. N. Y.: Marcell Dekker, 1983.
27. Roffael, Edmone, *Formaldehydabgabe von Spanplatten und anderen Werkstoffen*. Stuttgart: DRW, 1982.
28. Meyer B., Hermanns K. Formaldehyde release from urea-formaldehyde bonded wood products. J. Appl. Polym. Sci.: Appl. Polym. Symp. 1984; **40**: 27-39.
29. Kreibich R.E. Exposure of glue lines to weather. J. Appl. Polym. Sci.: Appl. Polym. Symp. 1984; **40**: 1-17.
30. Marutzky R., Dix B. Formaldehyde release of tannin-bonded particleboards. J. Appl. Polym. Sci.: Appl. Polym. Symp. 1984; **40**: 49-57.
31. Brown S.K. Hydrolysis and durability of urea-formaldehyde foam insulation. Polym. Degradation and Stability 1990; **27**: 121-43.
32. Cheshkov VM, Zahariev G, Natova MM. On some probes of application of amino resins in wood working industry, Proceedings of the X Symposium on Adhesives in Wood-Working Industry; 1991 September 11 - 13; Zvolen, 1991 (In Russian).
33. Simeonov J., Tcheshkov V., Zahariev G. Verfahren zur Härtung von Karbamidformaldehydharz. BRDPat., No. 2650265, 1979.
34. Simeonov J., Tcheshkov V., Zahariev G. Improvements in or relating to the curing of aminoplastic plastic materials. UK Pat., No. 1499285, 1978.

35. Simeonov J., Tcheshkov V., Zahariev G. Method of obtaining composite materials and products, particularly for construction and building purposes, based on urea-formaldehyde resins by addition of phosphogypsum. US Pat. No. 4 248 760.
36. Cheshkov V., Zahariev G., Rangelov V., Rusev D. Building elements from wooden particles. Proceedings of XI Symposium on Adhesives in Wood-Working Industry; 1993 September 8 -9; Zvolen, 1993 (In Russian).
37. Simeonov J., Rangelov V., Tcheshkov V., Zahariev G., Ivanov V. Gemisch für Holzflächen. DDR Pat. No. 272577, 1989.
38. Simeonov J., Rangelov V., Tcheshkov V., Zahariev G., Ivanov V. Anyagösszetétel préselt lapoknoz. Hung. Pat.No.192158, 1987.
39. Simeonov J., Rangelov V., Tcheshkov V., Zahariev G., Ivanov V. Smes k výrobe drevitýeh desek. CSR Pat. No.251 664, 1988.
40. Filyanov E.M. Influence of filler on glass-transition temperature of epoxy polymer and its relations with the properties of filled polymer. Vysokomol. Soedin. 1978; **20A**: 1845-48 (In Russian).
41. Mikhailova Z.V., Pugachevskaya N.F., Kovriga V.V. "Filled composite on the basis of unsaturated polyesters." In *Chemistry and Technology of High Molecular Compaunds*, A. Askadskii, ed. vol. 14, Moscow: VINITI, 1981.

CONCLUSIONS

One may adopt an acid-base point of view when tackling most of the interfacial phenomena that take place in filled polymers. As shown so far, the acid-base approach seems useful, since all fillers may be considered in practice as solid acids or solid bases. Moreover, a number of monomers, oligomers and polymers may be treated as compounds that are capable of displaying acidic or basic properties. But yet, the role of the acid-base interactions that occur during polymer filling is far from being entirely clarified. This gap is more or less due to the fact that researchers rarely characterize fillers by their acid-base properties and their role in the interfacial phenomena and interactions.

There are numerous methods of determining the acidity and/or basicity of inorganic compounds. Those methods are employed mainly when the compounds serve as catalysts of various chemical reactions, particularly in the petroleum processing industry. Depending on the methods employed, one may obtain different scales of the acidic or basic strength of the inorganic compound surface sites. Formulation of a correlation between the nature, strength and concentration of the acid-base sites and their activity during polymer filling is complicated. Moreover, polymer or composite formation in the presence of a filler often proceeds at high temperature, while the determination of the degree of filler acidity/basicity is usually carried out at room temperature. Thus, it is hard to interpret data on the adsorption interactions that occur in filled polymers and on the mechanism of compound formation on the filler surface.

We have shown that there are systems consisting of fillers and low-molecular analogues of polymers which adequately model adsorption processes. The acid-base interactions prove to be most important in the introduction of fillers during the stage of polymer synthesis. The adsorption interactions between the filler and the monomer or oligomer determine not only the properties of the final polymer, but also the kinetics and mechanism of polymer formation. In other words, the filler plays a two-fold role – as an active filler in forming a structural compound, and as a catalyst of the polymerization process. This role of the filler is most apparent when oligomers and monomers are filled with solid acids containing groups with strong basic-in-character properties (carbonyl, amide, etc. groups). In this way filler-catalysts exert control over distinct structural details of the forming polymer network that have important effects on the properties of the polymer composite. Knowing the acid-base properties of the composition components, we can partially predict the filler catalytic effect on the polymerization process. A correct choice of the filler-catalyst may enable one to change the structure formation and properties in a desired direction, i. e. to control not only the structure, but also the properties of the polymer that is formed.

Hence, we may successfully apply the acid-base concept to polymer filling. The mechanism of polymer structure formation, stimulated by adding active fillers (solid acids and/or solid bases), may serve as a basis for the appropriate filler choice. This knowledge on the polymer structure formation may also enable one to predict the filler catalytic performance during polycondensation and polymerization. Finally, one should account for the mechanism of structure formation when designing new composite materials and technological processes for the composite preparation and practical use.

We have shown that interfacial phenomena and processes are extremely complex, and that one can hardly find quantitative or qualitative agreement between the results of different authors. Hence, the experimental set-up and the sample history determination should be carefully chosen using more than one measurement technique. It is desirable to find compatibility between results of different authors before drawing any general conclusions.

There is no doubt that liquid polymer composites display irreversible structural breakdown under shear. However, for shear rates lower than a critical one that yields breakdown, polymers display Bingham behaviour. Those polymer characteristics were first reported fifty years ago. Later works confirmed the evidence but also introduced some uncertainties in assessing the properties of polymer composites. There are several important questions that one should answer before making further progress:

- What is the cause for the wide variation of the reported data in the yield value and plastic viscosity of composites that are characterized by acid-base interactions?
- Under exactly what conditions do interfacial phenomena and processes occur?
- How can one explain some anomalous changes of the material properties under continuous acid-base interactions that take place within the polymer composites?

The specific features of the rheological and mechanical behaviour of polymer composites, discussed so far, certainly do not encompass the whole available experimental material. Nevertheless, our approach not only substantially complements and explains the concepts developed in a number of works of that kind, but also affords definite generalizations. These include single-type filled polymer systems that model physicochemical processes and material physical and mechanical behaviour in the fluid state, highly elastic state; around the glass transition point, melting temperature, etc. The complex interrelation between physico-chemical interfacial phenomena, rheological properties and strength behaviour that one can attempt to trace is exceptionally important for an understanding of the general nature of polymer behaviour under changing temperature and mechanical conditions. Knowledge on those interrelated characteristics may produce sound principles of the study of composites and enable one to design an effective method of measurement and control, escaping from the limitations of the *ad hoc* empirical assessment of material properties.

Afterwords, we may hope for a uniform theoretical approach to polymer composites, enabling us to describe interfacial phenomena and various forms of material physico-chemical behaviour, starting from the processes that take place at the interface in the liquid stage and ending with the fabrication and processing of a solid polymer composite.

Index

GPSR Compliance

The European Union's (EU) General Product Safety Regulation (GPSR) is a set of rules that requires consumer products to be safe and our obligations to ensure this.

If you have any concerns about our products, you can contact us on ProductSafety@springernature.com

In case Publisher is established outside the EU, the EU authorized representative is:

Springer Nature Customer Service Center GmbH
Europaplatz 3
69115 Heidelberg, Germany

Batch number: 10583998

Printed by Printforce, the Netherlands